[*Notified in A.C.Is. 5th June, 1943.*]

> **NOT TO BE PUBLISHED**
>
> The information given in this document is not to be communicated, either directly or indirectly, to the Press or to any person not holding an official position in His Majesty's Service.

HANDBOOK OF ENEMY AMMUNITION

PAMPHLET No. 7

ITALIAN FUZES, GAINES, SHELL, CARTRIDGES, PRIMERS AND DETAILS OF SHELL MARKINGS GERMAN PRIMERS, SMALL ARM AND GUN AMMUNITION

By Command of the Army Council,

THE WAR OFFICE,
 5th June, 1943.

CONTENTS

Italian Ammunition

	PAGE
Markings on A.P., H.E., Hollow Charge and Shrapnel Shell	3
D.A./Graze Fuze with Clockwork Safety Device	6
Time Fuze Graduated to 13.2	8
Time Fuze Graduated to 160	12
Inneschi Graze Fuze	16
Gaines for H.E. and Piercing Shell	18
Primer, Percussion, Q.F., Cartridge, Model 35	20
47/32 Cartridge, Q.F., H.E. with Fuze	22
47/32 Cartridge, Q.F., A.P. with Tracer Fuze	26
47/32 Cartridge, Q.F., A.P.B.C. with Tracer Fuze	31
65/17 Cartridge, Q.F., H.E. with Fuze	31
75/27 Cartridge, Q.F. (Separate Loading)	33
75/27 H.E. Streamlined Shell, Model 32	35
75/13 Cartridge, Q.F. (Separate Loading)	37
75/13/27 H.E. Shell	37
100/17 Cartridge, Q.F. (Separate Loading)	39
100/17 H.E. Streamlined Shell	39
100/17 Hollow Charge Shell with Internal Base Fuze	41

German Ammunition

Primer, Percussion, Q.F. Cartridge C/13nA	44
Primer, Percussion, Q.F. Cartridge C/33	45
2 cm. Solothurn S.A. Cartridge with A.P./T. Shot (Tungsten Carbide Core) (2 cm. Pzgr. 40)	45
3.7 cm. Pak Cartridge, Q.F., A.P. with Tracer Fuze Bd.Z. 5103 or 5103*	47
7.5 cm. Pak 40 Cartridge, Q.F., A.P.C.B.C. with Tracer Fuze Bd.Z. 5103* (Panzergranate 39)	50
7.5 cm. Pak 40 Cartridge, Q.F., H.E. with Fuze Kl.A.Z.23 (Sprenggranate 34)	53
7.5 cm. Kw.K. 40 and Stu K 40 Cartridge, Q.F., A.P.C.B.C.	54
7.5 cm. Kw.K. 40 Cartridge, Q.F., H.E.	54
10.5 cm. l.F.H. 18 Q.F. Cartridges	55

ITALIAN MARKINGS ON A.P., H.E., HOLLOW CHARGE AND SHRAPNEL SHELL (Fig. 1)

The following markings have been met with in the course of examining captured Italian shell of the above-mentioned types. The types bearing these markings are of recent manufacture. The markings of earlier types do not always conform to the system.

Basic Colours of Body

With each of the types the body of the shell is pale blue (almost a light grey) and the head (*i.e.*, from nose to just above the shoulder) is red. When a cap is fitted to the shell the cap is red. The 100/17 hollow charge shell is an exception in that the cap is orange. Whether this is by design or is due to a variation in the shade of colour is not yet clear. These basic colours are normally applied by a process similar to sherardizing instead of by painting. The result is a flat finish and the absence of an obvious coating.

Bands

A green band immediately above the driving band is found on shell both of the piercing type and normal nose fuzed H.E. type when filled T.N.T. or amatol.

A brown band immediately above the driving band is found on shrapnel shell.

A white band appears to be the distinguishing marking of shell designed for the attack of armour. With armour piercing shell the band is at the approximate centre between the shoulder and the driving band. With hollow charge shell the band is immediately above the driving band.

Stencilling

The following particulars regarding the shell are normally stencilled in black between the shoulder and the driving band:—

(*a*) Weight of filled shell in kilograms.

(*b*) Nature of bursting charge. Shell filled with cast T.N.T. either by the pouring process or in the form of blocks are stencilled "TRITOLO." Those filled cyclonite/T.N.T. are stencilled "TRITOLITE." Shell filled amatol are stencilled "Amatolo."

(*c*) Letters indicating the filling factory followed by the last three figures of the year of filling.

(*d*) The calibre of the equipment in millimetres followed by an oblique stroke and the length of the piece expressed in terms of calibres.

The marking "Migl" indicates a modified design.

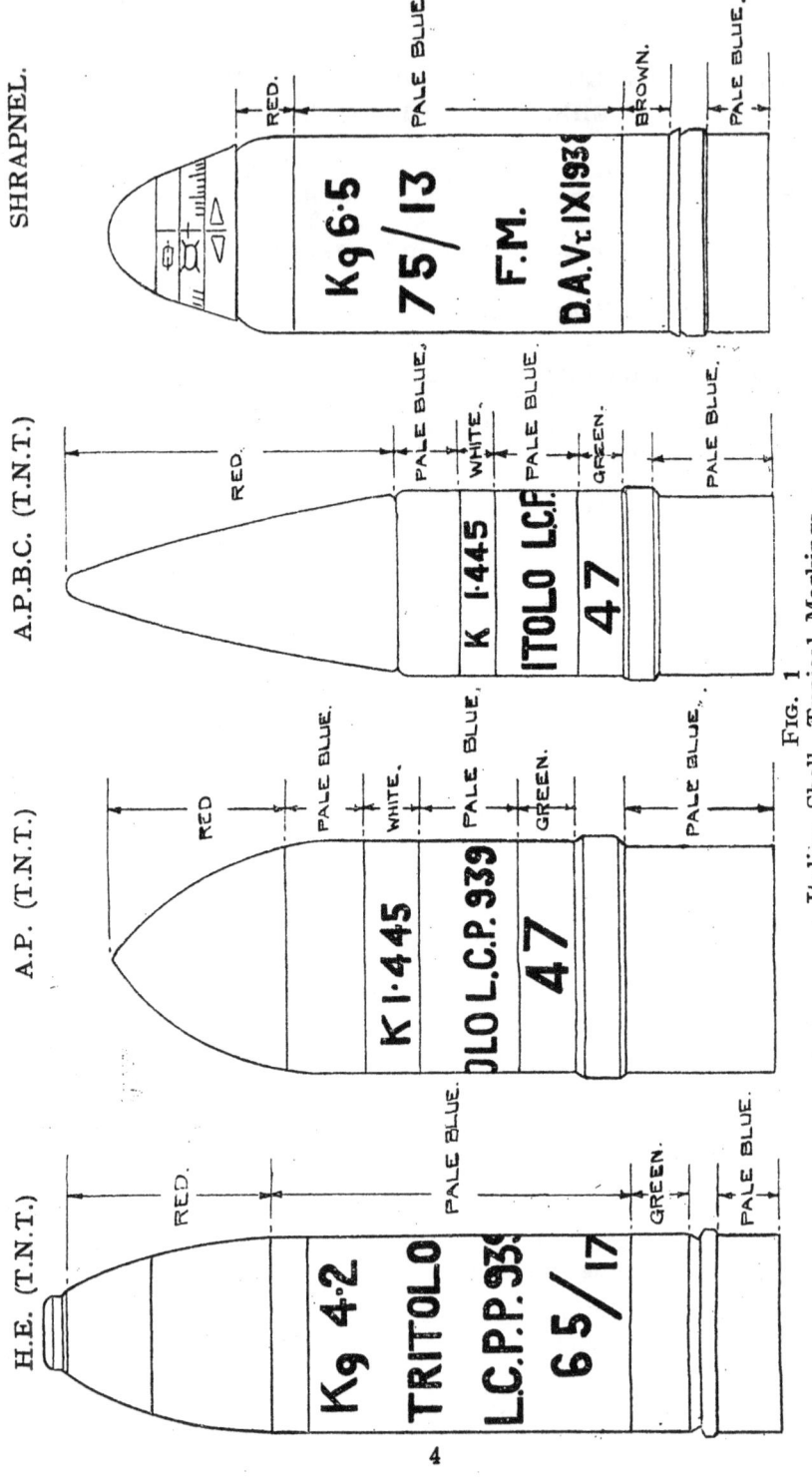

Fig. 1
Italian Shell, Typical Markings
(For 100/17 Hollow Charge Shell, see Fig. 19)

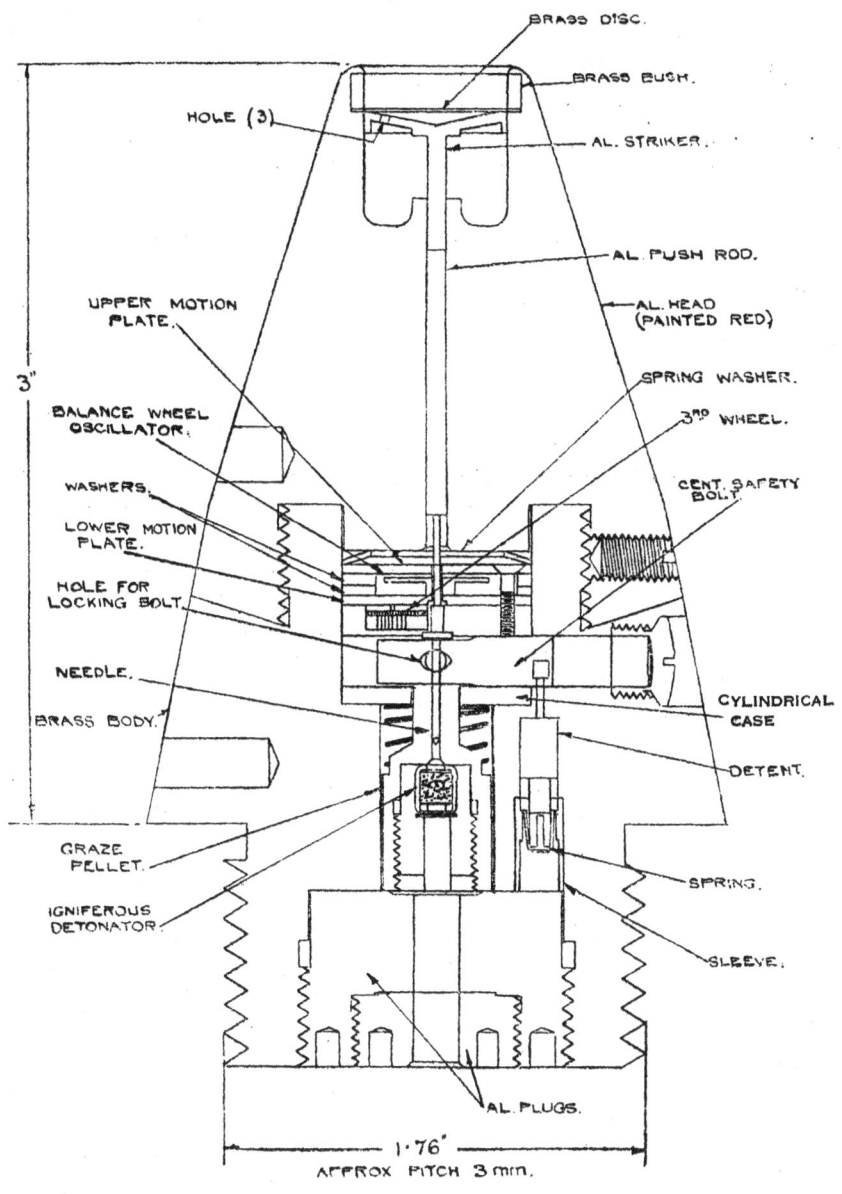

Fig. 2

Italian D.A./Graze Fuze with Clockwork Safety Device

ITALIAN D.A./GRAZE FUZE WITH CLOCKWORK SAFETY DEVICE (Figs. 2 and 3)

The fuze is of the floating needle type with a graze pellet carrying the detonator. The needle and detonator are held apart by a centrifugal safety bolt, the outward movement of which is controlled by a clockwork escapement. The design is similar in principle to the Model 16 described in Pamphlet No. 4, but differs in details.

The overall length of the fuze is 4 inches. When assembled in the shell the 3 inches which protrude consist of a brass tapering body with a red painted aluminium head. The head is tapered near its base to correspond with the shape of the body, but the taper is markedly increased further forward. The nose is flat and is fitted with a brass sealing disc.

The screw-threaded portion of the body for insertion in the shell is 1.76 inches in diameter and has a pitch of approximately 3 mm. The body is bored to accommodate the clockwork and graze mechanisms and is closed at the base by two aluminium screwed plugs which are drilled to provide a flash channel. The front end of the body is reduced in diameter and threaded for the assembly of the head. A lateral channel is formed for the safety bolt, and a recess, displaced from the centre, contains a detent and sleeve. A radial channel at right angles to the lateral channel is formed for the locking bolt.

The brass graze pellet carries a 2.2 grain igniferous detonator secured by a perforated brass plug and is designed with a necked portion which acts as a guide for the needle. A steel creep spring is held in compression between a shoulder on the pellet and the base of the brass cylinder containing the clockwork mechanism. The base of the cylinder is bored to fit over the neck of the graze pellet. The detonator filling consists of mercury fulminate 45.2 per cent., potassium chlorate 28.9 per cent. and antimony sulphide 25.9 per cent.

The brass detent is supported by a three-pronged form of stirrup spring which is attached to its base and rests on the shoulder formed in the brass sleeve. The stem of the detent enters a recess in the underside of the safety bolt and retains the bolt in the safe position until " set back " occurs.

The aluminium centrifugal safety bolt is forked at the inner end to provide two arms which pass under a flange formed on the needle and so prevent the needle moving towards the detonator. A recess to engage the detent is formed in the underside of the bolt and a hole is formed in the centre at one side to engage the stem of a spring-loaded centrifugal locking bolt. The opposite side of the safety bolt is in the form of a toothed rack which is enmeshed with the first spur of the clockwork escapement mechanism.

The clockwork arrangement is carried in a brass cylinder which is bored to accommodate the safety bolt and locking bolt. The escapement mechanism consists of a train of four wheels, each consisting of a spur and pinion, and a balance wheel oscillator. The first spur is enmeshed with the rack on the safety bolt. The fourth pinion is an escapement wheel and engages a recess formed in the eccentric projection on the balance wheel oscillator. The balance wheel oscillator is contained between the upper and lower motion plates which are suitably

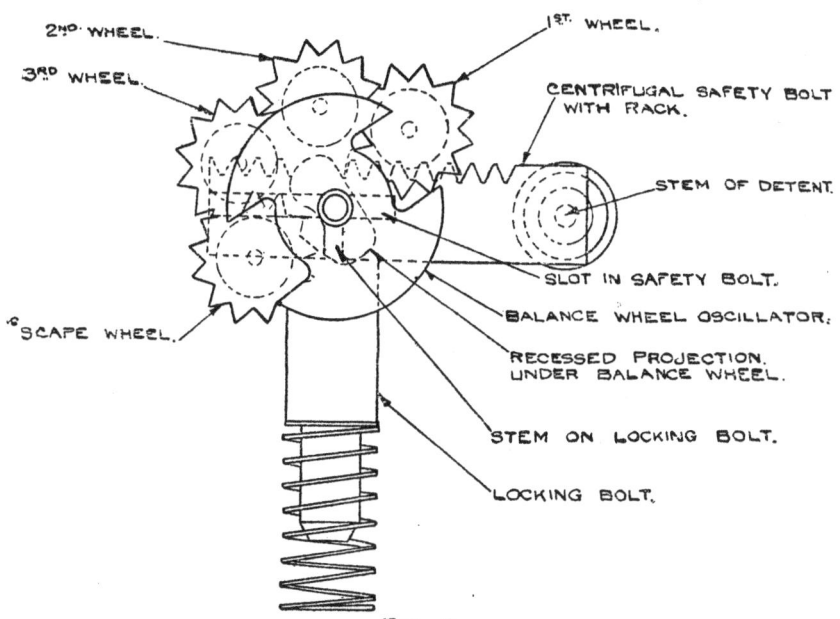

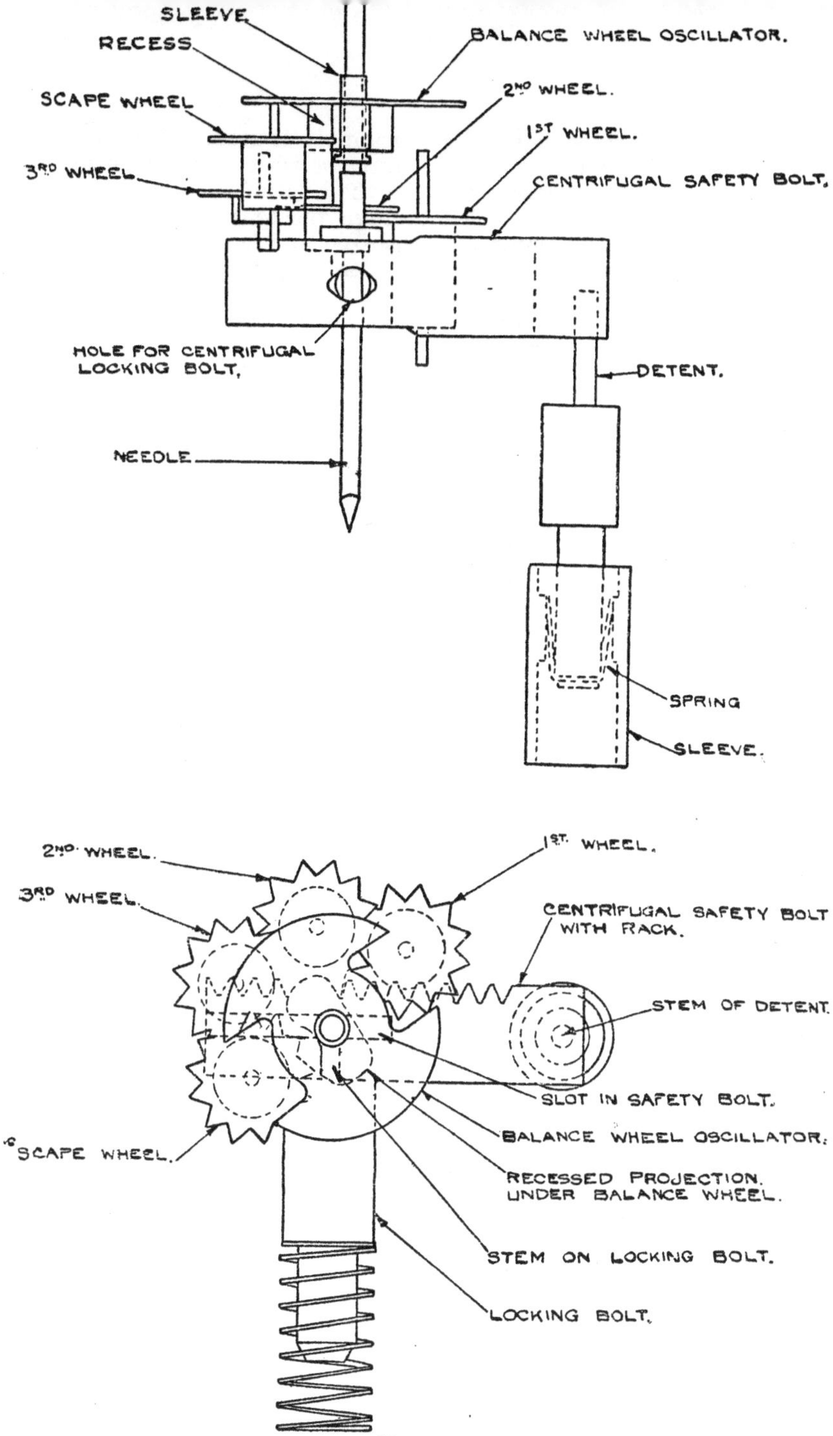

Fig. 3
Italian D.A./Graze Fuze. Arrangement of Mechanism

spaced by two brass washers. This assembly is secured by two screws to the brass cylinder containing the train of wheels and is surmounted by a spring washer compressed beneath the head of the fuze.

The aluminium head is drilled through the centre to take the stem of the hammer and the aluminium push rod and is recessed at the nose for the hammer head. The recess is closed against air pressure by a brass disc secured by a bush of the same material.

The hammer head has three holes for the escape of the air behind it when the hammer is driven in.

Action

On acceleration the detent sets back, forcing its spring past the shoulder in the sleeve and withdrawing its stem from the recess in the safety bolt. The prongs of the spring then engage the underside of the shoulder and prevent the detent moving forward.

During flight the locking bolt is moved outwards against its spring by centrifugal force and releases the safety bolt. The safety bolt then commences to move outwards, also under the impetus of centrifugal force. The outward movement of the safety bolt is transmitted by the toothed rack on the bolt to the train of wheels, and is controlled by the action of the balance wheel oscillator. The throw on the projection formed beneath the balance wheel oscillator, under the influence of centrifugal force, performs the normal function of the hair spring in a clock mechanism and, with the recess acting as a pallet, controls the rotation of the escapement pinion. When the forked portion of the safety bolt has moved clear of the needle and graze pellet, the needle is held off the detonator by the " creep " resulting from deceleration, and the graze pellet is held by the creep spring.

On graze the needle is driven in by the hammer and push rod whilst the graze pellet overcomes the creep spring and carries the detonator on to the needle. The flash passes through the flash channel in the base of the fuze.

ITALIAN TIME FUZE GRADUATED TO 13.2 (Fig. 4)

The fuze is used in the Q.F. 102 mm. 35 calibre anti-aircraft gun ammunition and is a tensioned fuze of the combustion type. The setting graduations extend from zero to 13.2. A fuze set to 13 gave a time of burning of 26.6 secs. at rest. The design includes a delay arrangement between the lower time ring and the magazine which ensures a minimum time of burning of .6 sec. and thus provides for safety against a " flash-over " in the bore or the results of a dangerously short setting. The screw-threaded portion of the fuze for insertion in the shell is 1.762 inches in diameter.

The exterior of the fuze, visible when assembled in the shell, is of aluminium or aluminium alloy. The tension cap and the upper ring are shaped to coincide and form an ogival head. The lower ring is cylindrical and the flange of the platform is tapered. The setting graduations may be marked on the lower ring or on the flange of the platform. A soldered alloy cover with tear-off wire and ring is sometimes fitted to the fuze. The cover is marked with a red arrow and the word " TIRARE," indicating the method of removal.

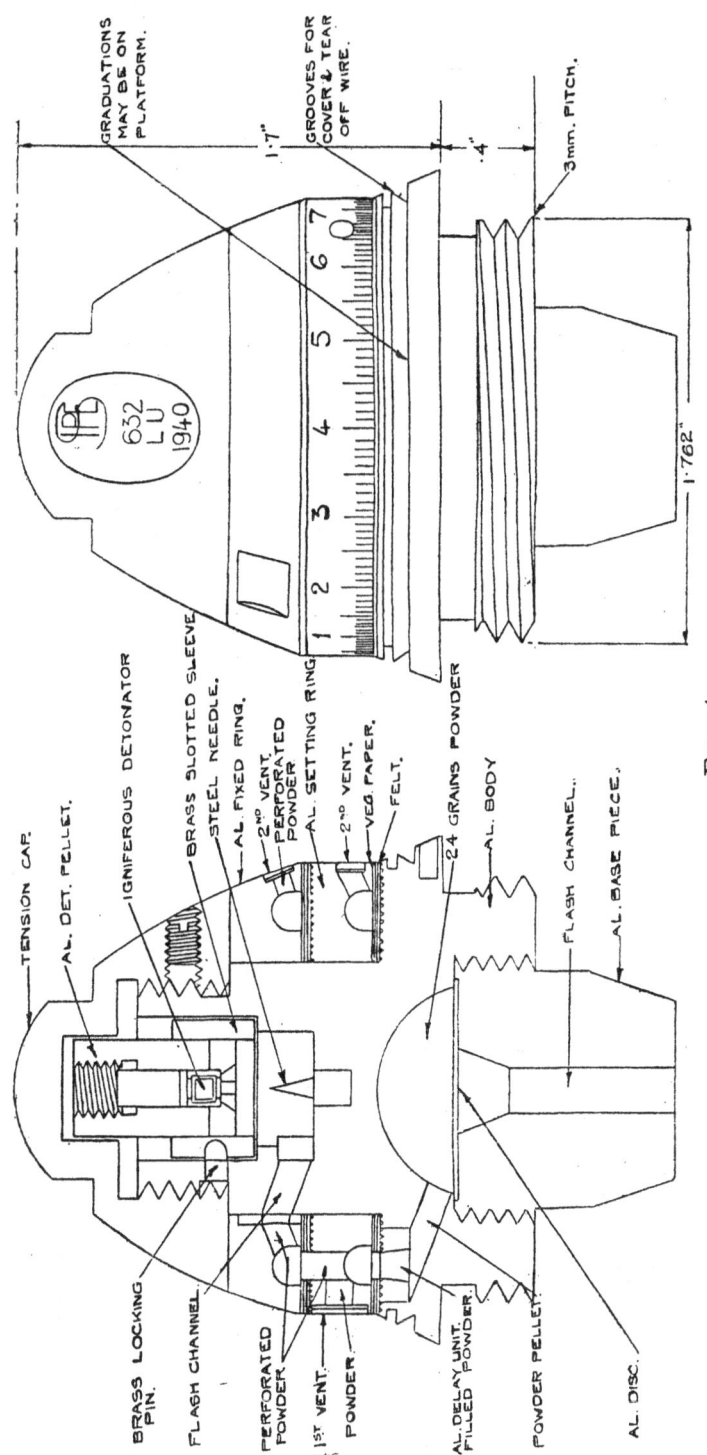

FIG. 4
Italian Time Fuze Graduated to 13.2

The aluminium body of the fuze is screw-threaded to a depth of .4 inch for insertion in the shell and is shaped to form a platform to support the lower time ring. The tapered flange of the platform may be grooved to receive the soldered alloy cover and the tear-off wire, or may be graduated for the setting of the lower time ring. The numbered graduations extend from 0 to 13. These are subdivided in tenths except between 0 and 1, where the subdivision commences at .8. The space between 0 and this graduation is barred out by two crossed diagonal lines. The subdivision continues beyond the 13 graduation to 13.2. A circular recess in the platform communicates with an inclined flash channel leading to the magazine. The recess contains an aluminium disc, with a central hole tapering towards the top, which contains a pressed filling of powder and introduces a .3 sec. delay. A felt washer, with a hole to correspond with the position of the delay unit, is attached to the platform by an adhesive. The inclined flash channel contains a solid pellet of powder, which gives a further delay of .3 sec. The magazine contains a 24-grain filling of granular gunpowder and is closed by an aluminium base piece which screws into the body with a left-hand thread and protrudes at the base. The base piece has a central flash channel, which is lightly closed at its inner end by a thin disc of aluminium. The body is cylindrical above the platform for the assembly of the two time rings, and is screw-threaded to receive the tensioning cap. Two semi-circular recesses are formed in the cylindrical part of the body which coincide with similar recesses in the upper time ring to receive the two aluminium locking pins which hold the upper time ring in a fixed position. A recess formed in the front end of the body contains the detonator assembly. A steel needle is fixed in the base of the recess, and a flash channel is drilled through the wall to coincide with a similar channel in the upper time ring.

The detonator assembly consists of an igniferous detonator carried in a cylindrical aluminium pellet which is held off the needle by a slotted brass sleeve. The detonator is retained in the pellet by a brass screwed plug. The slotted brass sleeve supporting the pellet is a tight fit round the base of the pellet, and is itself supported by a shoulder formed in the recess. Movement of the sleeve is prevented by a brass locking pin which is inserted through the wall of the recess and engages in the slot.

The lower time ring is of aluminium and is the setting ring. The groove containing the fuze powder in its underside extends through 310 degrees and has two circular vents for the escape of pressure and a flash-hole leading to the upper surface of the ring. The first vent and the flash-hole are connected with the commencement of the powder groove. The second vent is smaller and is located at 150 degrees from the commencement of the powder groove. The first vent contains a pressing of powder, and both vents are closed by tin discs, which are secured by stabbing and sealed with varnish. The flash-hole connecting the commencement of the powder groove with the upper surface of the ring contains a filling of pressed powder. A hole is drilled through this pressing and the powder in the groove beneath it to assist ignition. A washer of vegetable paper, with a hole to correspond with the commencement of the groove, is fixed to the underside of the ring to cover the powder groove, and a felt washer, perforated to correspond with the flash-hole, is fixed to the upper surface. A recess is formed in the cylindrical exterior wall of the ring for the setting key.

The fixed upper time ring is also of aluminium with a powder groove in its underside, which also extends through 310 degrees. Two vents are provided for the escape of pressure. The first is of the elongated type, and is formed in the bridge, or solid portion of the ring between the ends of the powder groove, and is connected to the commencement of the groove by a circular hole. This vent contains no filling, and is closed by a tin disc. The second vent is circular and smaller, and is located at 150 degrees from the commencement of the powder groove. A small hole through the powder filling is continued slightly into the fuze powder in the groove. The vent is closed by two tin discs. In both vents the closing discs are secured by stabbing and sealed with varnish. An inclined flash-hole is formed in the inner wall of the ring at the commencement of the powder groove. The powder filling in the flash-hole has a central perforation, which corresponds with a similar vertical perforation in the fuze powder at .1 inch from the commencement of the groove. The flash-hole in the ring is held coincident with the flash-hole in the body by the two locking pins. A washer of vegetable paper is fixed to the underside of the ring to cover the powder groove. A hole in the washer corresponds with the commencement of the groove.

The aluminium tension cap is screwed over the forward end of the body to obtain the required resistance to the turning of the lower ring, and is secured by a brass grub screw.

Analysis of the powders gave the following results:—

	Fuze Powder.	Magazine Powder.
Sulphur	8.4 per cent.	9.5 per cent.
Potassium nitrate	73.2 ,,	75.5 ,,
Charcoal	14.7 ,,	14.6 ,,
Tarry matter	3.7 ,,	0.4 ,,

Action

On acceleration the detonator pellet sets back through the slotted sleeve and the detonator is impinged on the needle. The flash produced passes through the flash channel in the recess and ignites the powder in the flash channel in the upper ring, thus igniting the fuze powder, which commences to burn along the groove in the underside of the ring. The pressure set up by the burning fuze powder dislodges the closing disc of the first vent, and thus prevents variation in the rate of burning as the result of heat and pressure. The distance between the commencement of the powder groove in the upper ring and the powder-filled flash-hole in the surface of the lower ring depends upon the angle through which the lower ring is turned in setting. When the fuze powder in the upper ring has burned round to this surface flash-hole the fuze powder in the lower ring is ignited through the powder in the flash-hole. The closing disc of the vent is blown out and the fuze powder burns round the groove to the delay unit in the platform. From the delay unit the flash is passed to the solid powder pellet in the flash channel, and thence to the magazine.

The second vent in each of the rings is blown open when the fuze powder has burned round to the position of the vent.

With the lower ring set to the zero graduation the flash-hole in its surface is aligned with commencement of the powder groove in the upper ring and the delay unit in the platform. Thus the flash from the detonator is transmitted directly to the delay unit.

With the lower ring set to the cross which indicates the safe position, the surface flash-hole in the lower ring is masked by the solid portion between the ends of the groove in the upper ring. Also, the delay unit in the platform is masked by the corresponding portion of the lower ring.

ITALIAN TIME FUZE GRADUATED TO 160 (Figs. 5 and 6)

The fuze is used in Q.F. 102 mm. 35 calibre anti-aircraft gun ammunition and is a tensioned fuze of the combustion type. The screw-threaded portion of the body, for insertion in the shell, is longer than that of the smaller fuze graduated to 13.2. The setting graduations extend from zero to 160. A fuze set to 160 gave a time of burning of 35.9 secs. at rest. The design includes a delay arrangement between the lower time ring and the magazine, which ensures a minimum time of burning of .6 sec. as in the smaller fuze. The body is screw-threaded to a 1.766 inch diameter, with a pitch of approximately 3 mm.

With the exception of the graduated brass platform, the exposed part of the fuze when fitted in the shell is of aluminium. The nose plug has a flat head, and is shaped to correspond with the tapering contour of the tension nut and upper time ring. The lower time ring is cylindrical and is milled. The platform is tapered. A brass cover with slight fluting near the nose is attached by a soldered tear-off strip to the lower edge of the platform.

The brass body has a magazine, containing 51 grains of gunpowder, formed in its underside. The magazine is closed by a screwed brass plug which protrudes from the base of the fuze. The plug has a central flash channel closed by a thin brass disc. Six equi-spaced recesses, surrounding the flash channel, are drilled from the base of the plug. The plug is secured by means of a set screw. A central channel is formed leading from the magazine to the top of the body. The flash channel between the magazine and the platform is formed by a radial channel, closed at its outer end by a brass plug, and a vertical channel connecting the radial channel to the surface of the platform. The radial channel contains a solid pellet of powder, which causes a delay of .3 sec., and the vertical channel contains a brass ring with a hole tapering towards the head, which is filled with pressed powder, and also provides a delay of .3 sec. The central channel and the radial channel are closed in the magazine by discs of tinfoil. The numbered setting graduations on the flange of the platform extend from zero to 160 and are subdivided in tenths from .8 onwards. The .5 setting is also graduated. A radial locking pin of brass is fitted in the body to engage a recess in the fixed upper time ring. The body is screw-threaded near the forward end to receive the tensioning nut, and is reduced in diameter at the head to form a spigot which is threaded to receive the cover plate and two locking nuts. The step formed by the reduction in diameter at the head has formed in it a part-circular groove,

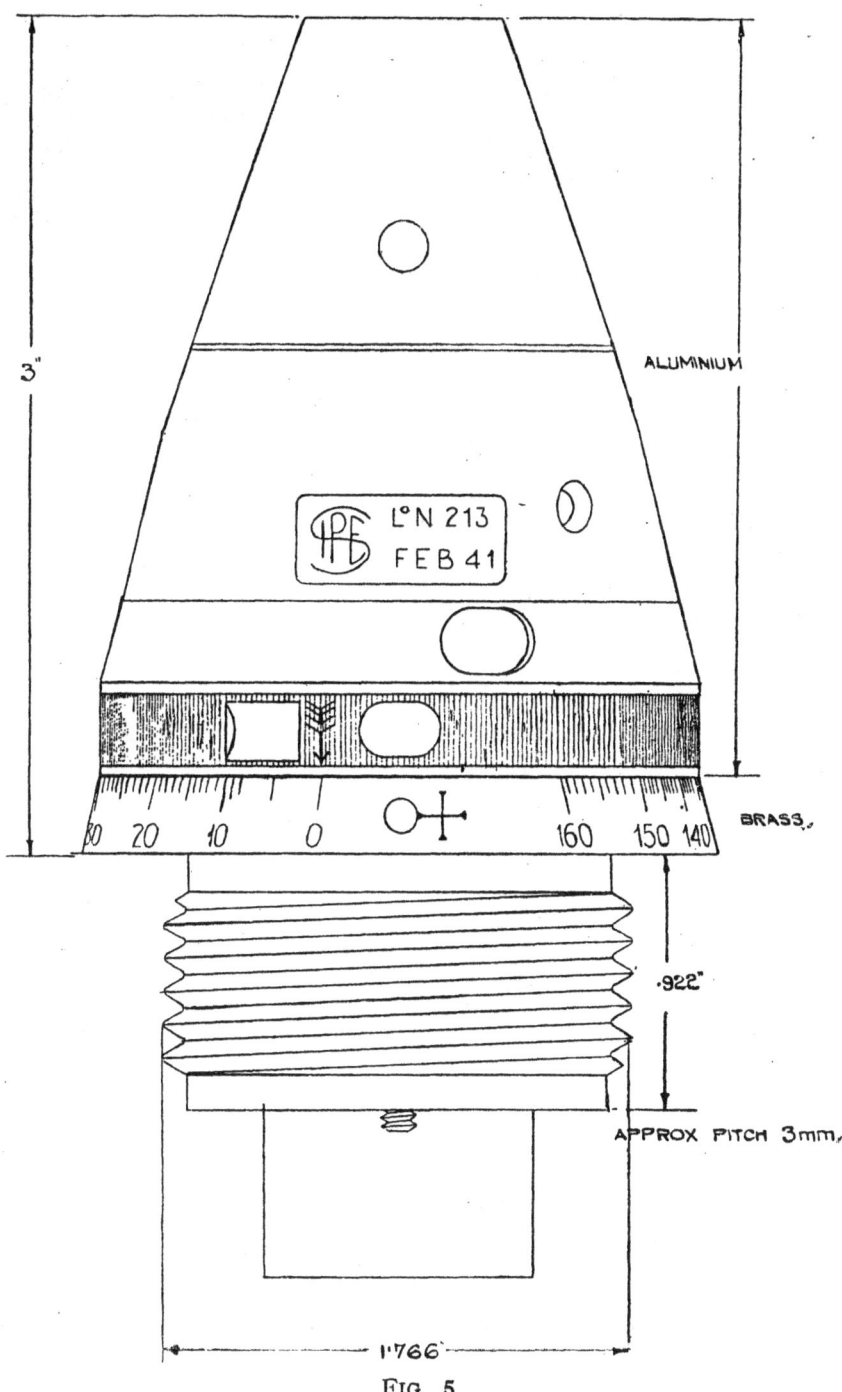

Fig. 5

Italian Time Fuze Graduated to 160

the ends of which are connected by inclined grooves with vertical grooves in the recess containing the detonator assembly. This recess is displaced from the centre of the body, and has a steel needle fitted in its base. Two inclined flats are formed at the front end of the recess, on which the arms of the stirrup spring are supported, and a flash channel for the ignition of the upper time ring is formed in the wall of the recess.

The detonator assembly consists of an igniferous detonator carried in a cylindrical brass pellet, which is held away from the needle by a brass stirrup spring.

The lower time ring is the setting ring, and has a powder groove in the underside extending through 320 degrees, with three vents. The first vent corresponds with the commencement of the groove, the second is at 115 degrees from this point, and the third at 225 degrees. An arrow is inscribed on the milled exterior of the ring for setting, and a recess is formed for the setting key.

The upper ring is locked by a radial brass pin in the body. The powder groove in its underside extends through the same angle as that in the lower ring, and has vents similarly placed except that the first vent is in the solid part of the ring and is connected to the groove by a short channel.

The other details of the rings are the same as those given in the description of the smaller fuze.

The brass cover plate consists of a disc with a screw-threaded hole in the centre, and is screwed on the spigot projection on the body to cover the detonator recess and the curved groove in the body.

The cover plate is held by a brass locking nut screwed to the spigot above it.

The tensioning nut is formed with a diaphragm perforated to fit over the spigot on the body, and is screwed over the body to bear on the time rings to produce the required resistance to the turning of the lower ring. After adjustment the tensioning nut is secured by a brass locking nut, which is screwed to the spigot and bears on the diaphragm, and by a fixing screw. The nut has an internal screw-thread and a fixing screw for the attachment of the nose plug.

The aluminium nose plug is tapered with a flat top and is screwed into the tensioning nut, with a lead washer sealing the joint.

Analysis of the powders gave the following results:—

	Fuze Powder.	Magazine Powder.
Sulphur	9.2 per cent.	9.6 per cent.
Potassium nitrate	75.0 ,,	75.5 ,,
Charcoal	15.3 ,,	14.5 ,,
Tarry matter	0.5 ,,	0.4 ,,

Action

On acceleration the detonator pellet overcomes the support of the stirrup spring and sets back, carrying the detonator on to the needle. The flash produced is transmitted to the filling in the upper time ring, and the subsequent action is similar to that of the smaller fuze. The groove in the detonator recess and in the front end of the body appear to be designed for the dispersion of the pressure set up by the action of the detonator, and thus avoids disintegration of the powder.

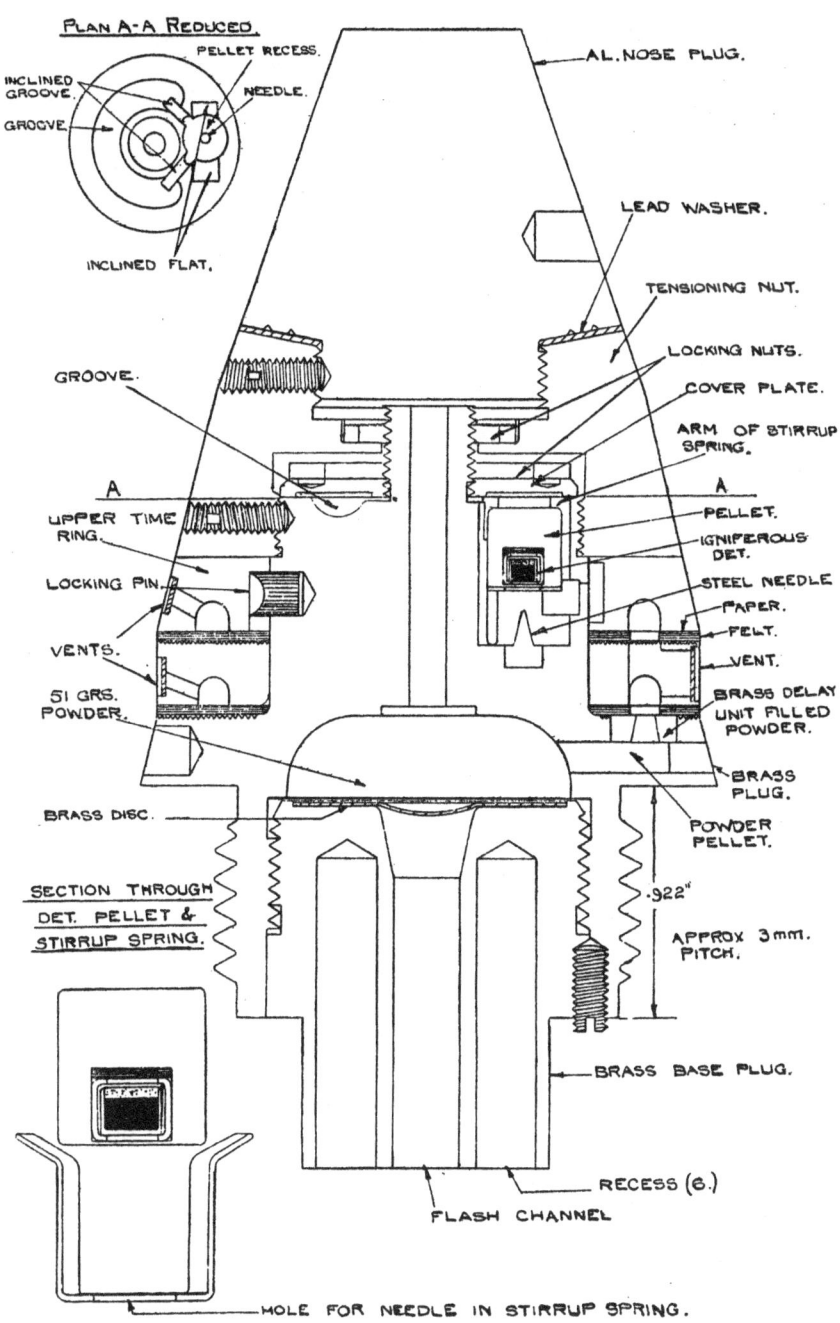

FIG. 6 .
Italian Time Fuze Graduated to 160

ITALIAN INNESCHI GRAZE FUZE (Fig. 7)

The fuze is used in H.E. shell for a number of equipments and is of the igniferous type, with a graze mechanism in which the detonator is fixed and the needle is carried in the graze pellet. The body is of .97 inch gauge, with approximately eleven threads to the inch. In order to obtain detonation, a gaine is inserted in the fuze-hole beneath the fuze, the following types having been met with:—

.89 gram cyclonite gaine
1.8 gram cyclonite gaine
2.3 gram ballistite gaine

For safety in transport and storage the screwed brass plug carrying the detonator is not inserted in the head of the fuze. Instead, an aluminium or black composition transit plug is fitted.

The cylindrical brass body is screw-threaded throughout its length for insertion in the shell, and has a flange at the head which is only part protruding from the fuzed shell. A screw-threaded hole in the head receives either the detonator plug or transit plug and leads to the interior recess containing the graze mechanism. The diameter of this recess is increased to form a shoulder near the base which locates the retaining collar, and below the shoulder it is screw-threaded to receive the magazine.

The brass arming sleeve has an internal circumferential groove to engage the prongs of the arming spring when in the armed position, and is fitted over the head of the graze pellet, where it is supported by the arming spring. In this position it prevents the needle moving towards the detonator.

The arming spring consists of a brass ring which fits over the forward end of the graze pellet and is shaped to form three prongs or arms which support the arming sleeve.

The brass graze pellet is in the form of a cup with two recesses cut diametrically opposite in the wall at the mouth to receive the needle bar, which is of steel, with a pyramid pointed needle formed at its centre. A flange formed round the base of the pellet is cut away at four equidistant positions, with a semi-circular chamfer on the underside. At each of these positions an inclined flash channel is drilled between the chamfered surface and the base of the interior of the pellet. A hole is drilled through the centre of the base to receive the stem of the sealing plug.

The copper sealing plug is shaped to seat in and seal the flash-hole of the magazine and is provided with a stem screw-threaded to two diameters for the assembly of the stirrup spring and attachment to the graze pellet. The stirrup spring is held between the head of the plug and a brass nut screwed on to the stem immediately behind the head. The upper part of the stem is inserted through the hole in the base of the graze pellet and secured by a brass nut inside the pellet.

The steel stirrup spring has two inclined arms which are engaged by the wall of a conical hole in the retaining collar.

The brass retaining collar consists of a disc, of larger diameter than the recess containing the graze pellet, with a tapered central hole to receive the stirrup spring. The collar bears against the shoulder in the body, and is secured by the magazine.

The brass magazine screws into the base end of the body, and contains a .8 grain perforated pellet of gunpowder. A flash-hole formed in

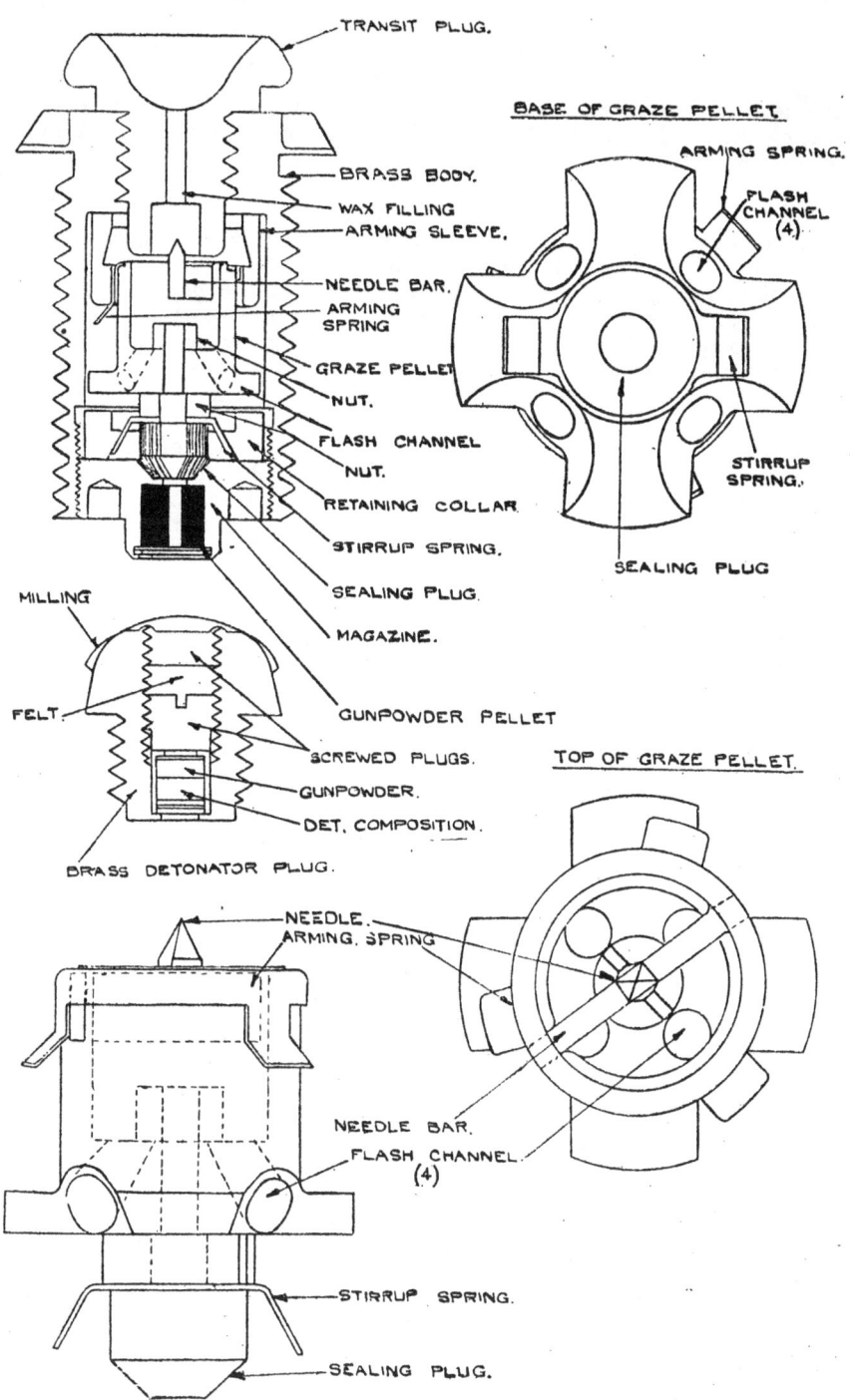

FIG. 7
Italian Inneschi Graze Fuze

the head is tapered and provided with a soft metal seating to receive the sealing plug. The base is lightly closed by a brass washer, with a fabric disc on its inner side.

Detonator Plug

The brass plug is mushroom shaped, with a screw-threaded stem for insertion in the head of the fuze and a ring of knurling on the head. A central recess for the detonator is formed in the plug with a flash-hole at its base and a screw-thread in the upper part to receive the closing plugs. The detonator is held in the lower part of the recess and has a copper shell closed at the top by a copper disc and at the bottom by a brass disc and white paper disc. The filling consists of a 1 grain pressing of gunpowder with a 1.2 grain sensitizing layer of detonating composition. The composition consists of mercury fulminate 12.4 per cent., potassium chlorate 45.3 per cent., antimony sulphide 39.1 per cent., and grit 3.2 per cent. The recess is closed by two brass screwed plugs, with felt packing inserted between them.

Action

On acceleration the sealing plug sets back in its seating and seals the flash channel, thus affording protection should the detonator fire prematurely. At the same time the arming sleeve sets back over the graze pellet and is retained in the set-back position by the prongs of the arming spring engaging in the internal groove in the sleeve. The graze pellet with its needle is then held off the detonator only by the arms of the stirrup spring bearing against the wall of the tapered hole in the retaining collar. On graze the momentum of the graze pellet forces the arms of the stirrup spring through the retaining collar and the pellet moves forward, thus causing the needle to pierce the detonator and removing the sealing plug from its seating. The flash from the detonator passes through the flash-holes in the graze pellet and enters the magazine. The explosion of the powder pellet in the magazine produces a flash which passes to the initiator filling in the gaine, and thus brings about the detonation of the gaine.

ITALIAN GAINES FOR H.E. AND PIERCING SHELL (Fig. 8)

The following types of gaines are used in conjunction with igniferous fuzes to obtain detonation of the bursting charge in H.E. shell and piercing shell:—

.89 Gram Cyclonite Gaine

The gaine is used with nose fuzes and base fuzes.

When used in conjunction with a base fuze the gaine is fitted to the fuze either by means of a screwed adapter (Fig. 8A) or by being screwed directly into the head of the fuze.

When the gaine is used with a nose fuze it may be screwed into the base of the fuze, or it may be fitted with a plain flanged adapter (Fig. 8B) and inserted in the mouth of the exploder container in the shell.

The filling of the gaine is contained in an aluminium capsule. 1.28 inches long and .29 inch in diameter. The capsule is open at one end,

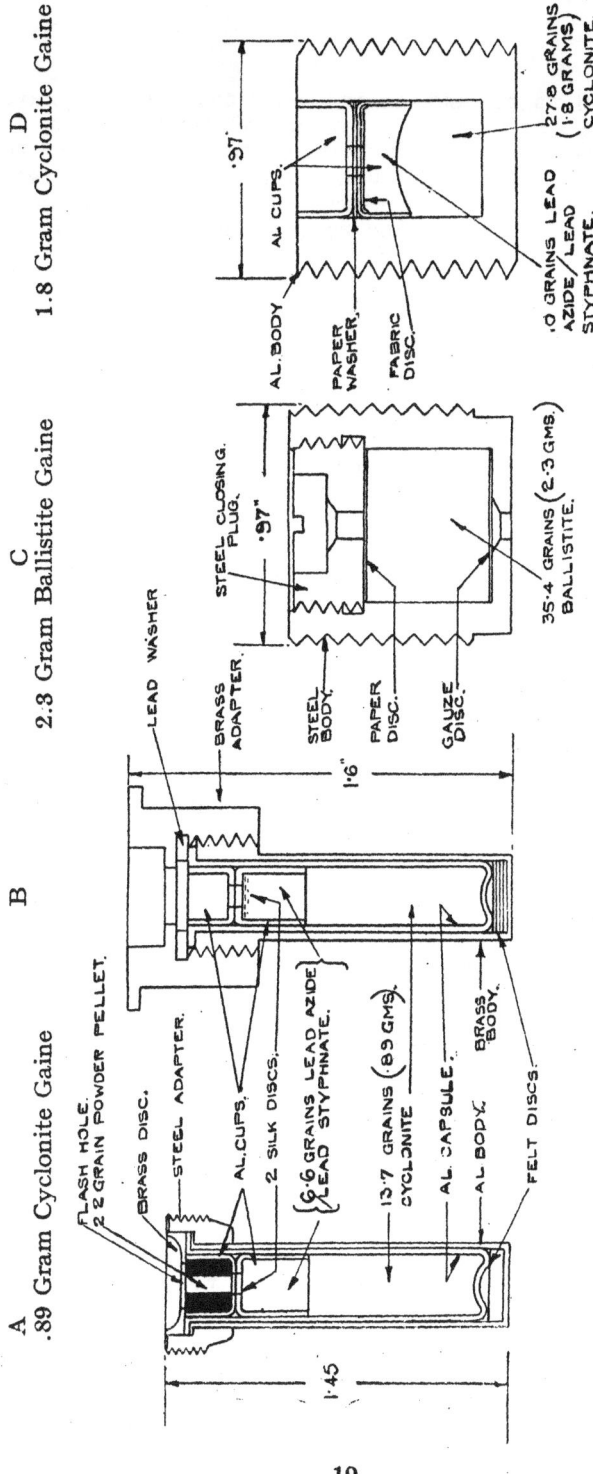

FIG. 8

Italian Gaines for H.E. and Piercing Shell

A .89 Gram Cyclonite Gaine
B
C 2.3 Gram Ballistite Gaine
D 1.8 Gram Cyclonite Gaine

with an external flange, and, at the other end, has a concave base. The main filling consists of 13.7 grains (.89 gram) of cyclonite, and above this is an initiator in the form of an inverted cup containing 6.6 grains of lead azide/lead styphnate (55 per cent./45 per cent.). The cup has a hole in the base which is lightly closed by two discs of white silk. The capsule is closed at the mouth by a perforated cup which in some instances contains a 2.2 grain perforated pellet of gunpowder.

2.3 Gram Ballistite Gaine (Fig. 8C)

The gaine is used in H.E. shell under the Inneschi graze fuze with a ballistite exploder. The screw-threaded steel body is .97 inch in diameter and has approximately 11 threads to the inch.

The filling consists of 35.4 grains (2.3 grams) of ballistite, and is contained in the cup-shaped interior of the gaine, which has a flash-hole at the base lightly closed by a gauze disc. The gaine is closed at the head by a steel screwed plug with central flash-hole. The hole is closed by a paper disc adhering to its underside. The ballistite contains approximately 45 per cent. of nitroglycerine and the nitrocellulose has a 12.3 per cent. nitrogen content.

1.8 Gram Cyclonite Gaine (Fig. 8D)

The gaine is used in H.E. shell under the Inneschi graze fuze with a T.N.T. exploder. The diameter and number of threads to the inch are the same as with the ballistite gaine.

The aluminium body is screw-threaded over the whole of its length and is solid at the base. The filling consists of 27.8 grains (1.8 grams) of cyclonite under an inverted aluminium cup containing 10 grains of lead azide/lead styphnate (40.1 per cent./59.9 per cent.). The base of the cup has a central flash-hole closed by a fabric disc. The gaine is closed by a similar cup, inserted base downwards, with a paper washer beneath it.

ITALIAN PRIMER, PERCUSSION, Q.F., CARTRIDGE, MODEL 35
(Fig. 9)

The primer body is of brass with a flange at the base, and is screw-threaded externally for insertion in the cartridge case. The diameter of the threaded portion is approximately .55 inch. The interior of the body has a small recess near the base to accommodate the initiator capsule and is screw-threaded to receive the magazine.

The initiator capsule is of copper and contains a brass anvil with a conical flash-hole and .86 grain of initiator composition. The composition consists of mercury fulminate 31.9 per cent., potassium chlorate 32.4 per cent., antimony sulphide 32.6 per cent., and ground glass 3.1 per cent. The composition is protected by varnish, but there is no foil disc.

The brass magazine is in the form of a cup with a flash-hole in the base and an external screw-thread for assembly in the body. The magazine contains a 14.2 grain pellet of gunpowder with a central perforation.

The primer is closed by a brass disc over which the mouth of the body is turned and the closing sealed with varnish.

Base Stampings

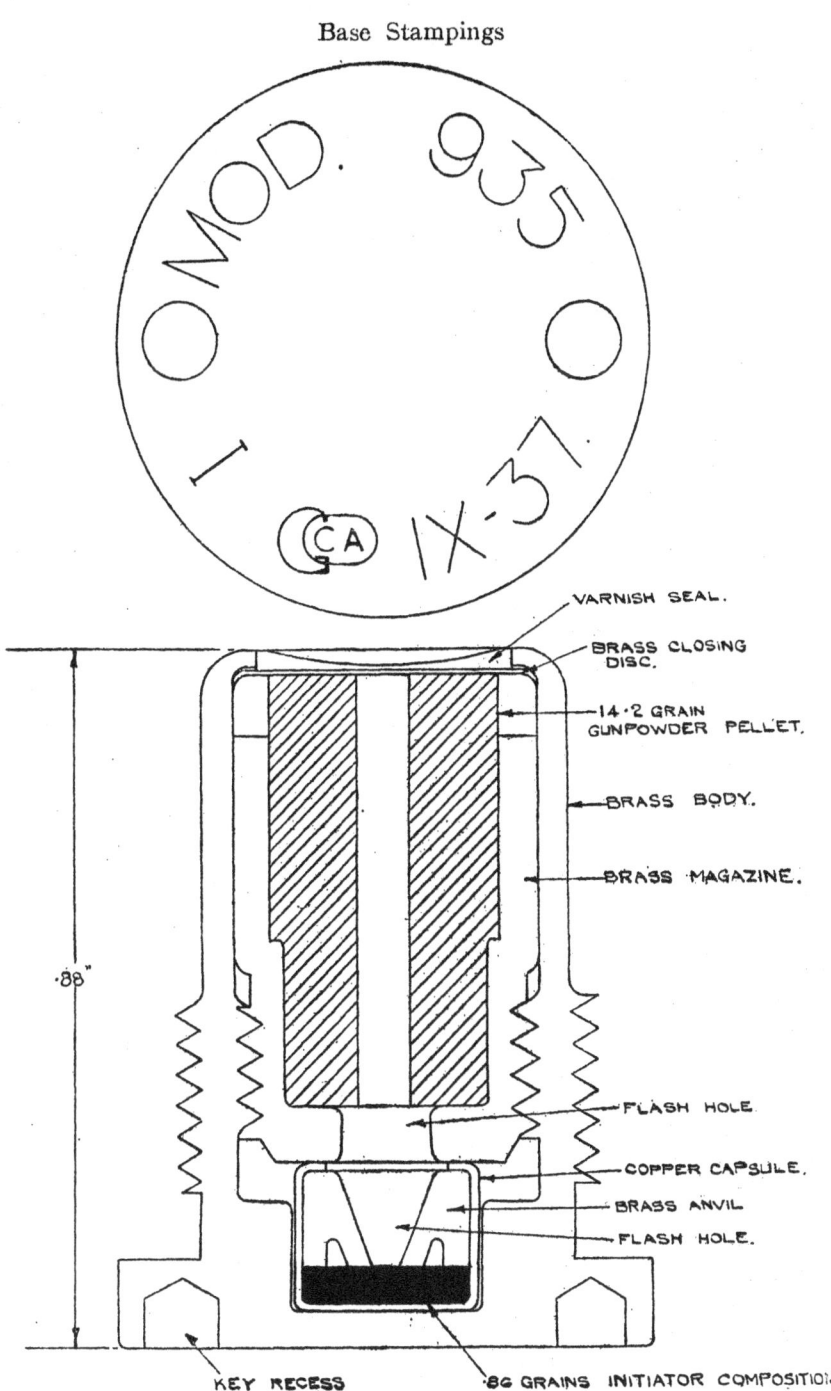

Fig. 9
Italian Primer, Percussion, Q.F. Cartridge, Model 35

Action

When the base of the primer is struck centrally the initiator composition is crushed on the anvil and initiated by mechanical shock. The flash produced by the composition passes through the flash-hole in the anvil and the magazine and ignites the powder pellet. The explosion of the powder pellet blows out the closing disc and ignites the propellant charge. The escape of propellant gases over the primer is prevented by the expansion of the body in the primer boss. Internal sealing is provided by the base of the primer.

ITALIAN 47/32 CARTRIDGE, Q.F., H.E. WITH FUZE (Fig. 10)

The fixed Q.F. round is fired from the 47 mm. 32 calibre anti-tank gun, model 35.

The weight of the complete round is approximately 6 lb. 4 oz. and the overall length 14.1 inches. The shell is coloured pale blue with a red head and a green band immediately in front of the driving band. The T.N.T. bursting charge is indicated by the stencilling " TRITOLO."

The complete round consists of the following components:—

    H.E. shell filled T.N.T.
    Combined graze and direct action fuze with gaine.
    Brass cartridge case.
    Propellant charge of double base composition.
    Percussion primer, model 35.

Shell

The shell, without the fuze, is approximately 9.2 inches in length, and weighs 5.2 lb. filled and fuzed. The ratio of length to diameter is abnormally high, and the copper driving band is more than a third of the length from the base. The shell is machined from rolled bar steel and has a parallel walled cavity of the same diameter as the fuze hole, and is thus suitable for block filling. The exterior of the shell is thinly coated with zinc before colouring. There is no base plate.

The weight of the empty shell is 4 lb. 11 oz.

Bursting Charge

The bursting charge consists of five pressed pellets of T.N.T. in a transparent paper wrapper. The uppermost pellet has a cavity to receive the gaine fitted to the fuze. A felt disc is inserted in the base of the shell cavity below the filling and a washer of the same material is inserted between the top of the filling and the fuze. The weight of the bursting charge is 5 oz. 3½ dr.

Fuze (Fig. 11)

The fuze has a combined graze and direct action, and is fitted with a .89 gram cyclonite gaine to obtain detonation.

The body of the fuze is screw-threaded externally for insertion in the shell and is prepared at the base to receive the gaine container. The threaded portion is of 1.14 inch diameter and the pitch approximately 2 mm. At the front end a screw-threaded recess is formed to receive the needle pellet holder, which also provides the means of attachment for the screwed head of the fuze. A large recess in the centre of the

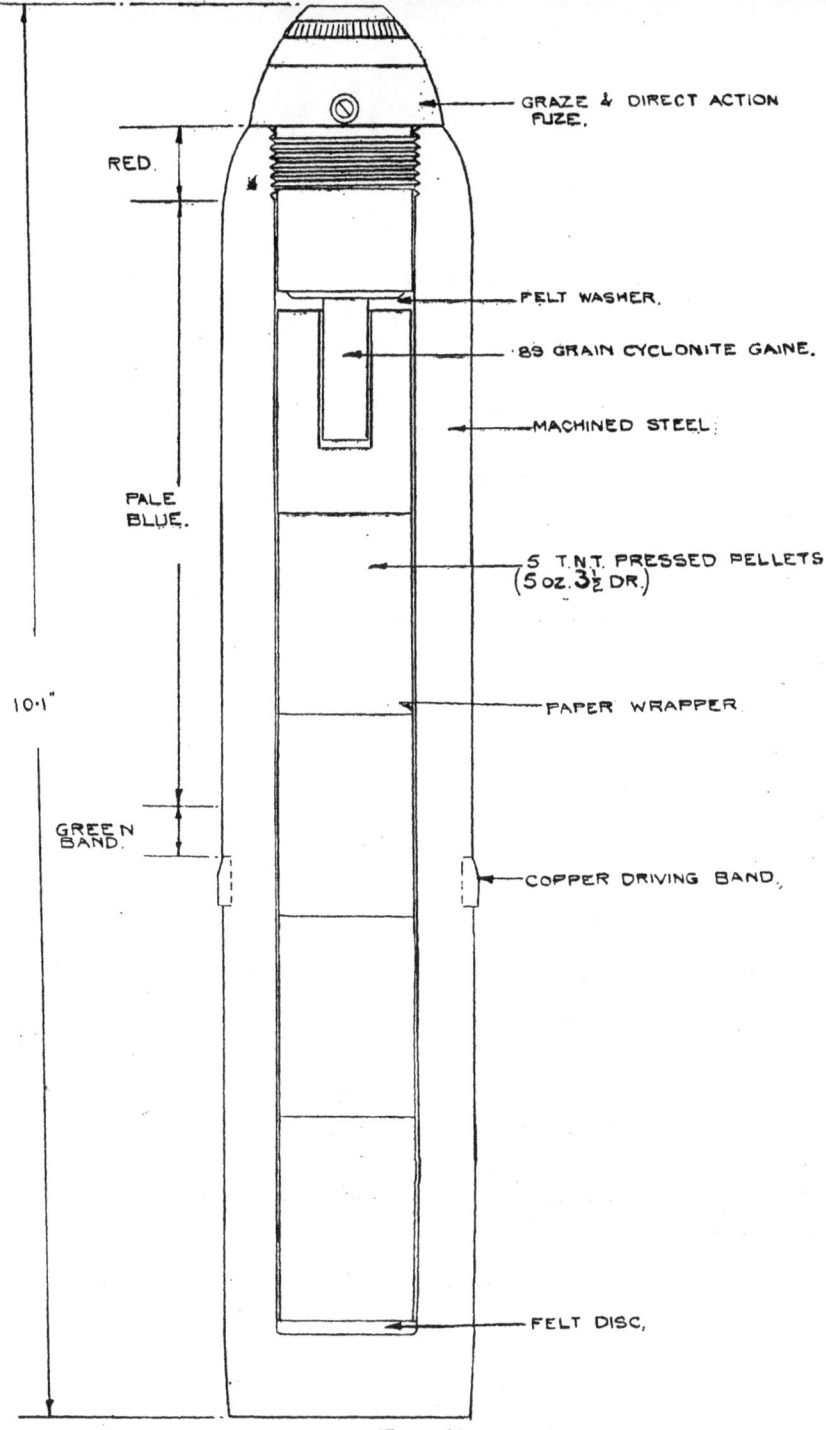

FIG. 10
Italian 47/32 Q.F., A/T Gun, Model 35, H.E. Shell with Fuze

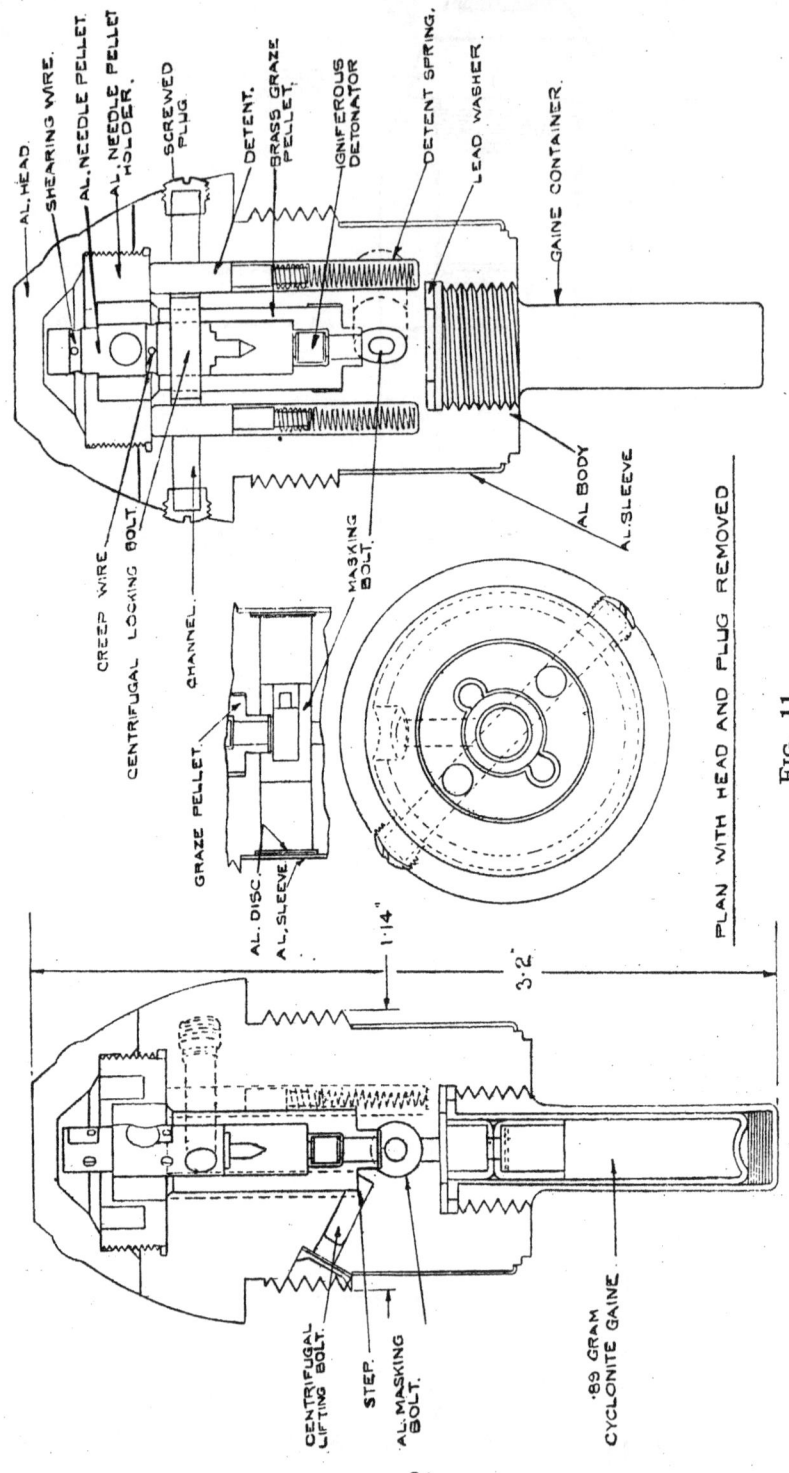

Fig. 11

Italian 47/32 H.E. Shell, Direct Action/Graze Fuze

body accommodates the graze pellet and has a flash channel at the base leading to the magazine. A lateral channel containing a masking bolt intersects the flash channel and is closed at both ends by aluminium discs. These discs are covered by an aluminium sleeve which is fitted round the body below the screw-threads. Another lateral channel is formed in the upper part of the body to accommodate the centrifugal locking bolt which passes through holes in the graze pellet and the needle pellet. The bolt is retained in position (so that its protruding ends engage in the channel on each side of the graze pellet) by two spring-loaded detents. The detents and springs are carried in recesses formed in the body. The lateral channel is intersected by these recesses and is closed at the ends by screwed plugs. An inclined channel, between the screw-threaded portion of the body and the recess for the graze pellet, contains a centrifugal bolt, which is stepped to engage the base of the graze pellet.

The needle pellet is fitted with a shearing wire near its front end. The protruding ends of the wire impede the passage of the pellet through the hole in the holder. A second wire with protruding ends is fitted in the needle pellet to engage the mouth of the cup-shaped graze pellet and thus prevent creep during deceleration in flight. A channel is formed in the needle pellet to coincide with the holes in the wall of the graze pellet and to receive the centrifugal locking bolt.

The graze pellet is in the form of a cup, the mouth of which fits over the lower end of the needle pellet. An igniferous detonator is fitted inside the base of the cup, and below this a flash channel passes through a spigot formed on the base. The spigot engages in a recess in the masking bolt and locks it in the masking position. The detonator consists of an inner and outer copper cylinder with a tinfoil lining and a copper closing disc at each end, and contains a .8 grain filling consisting of mercury fulminate 6.6 per cent., potassium chlorate 40.5 per cent., antimony sulphide 47.8 per cent., and grit 5.1 per cent.

The masking bolt consists of a hollow cylinder with a recess to engage the spigot at the base of the graze pellet.

The .89 gram cyclonite gaine fitted to the fuze is described in the section on Italian gaines in this pamphlet. In this instance the outer cup does not contain a powder pellet.

Action

On acceleration the two detents set back, leaving the upper lateral channel clear for subsequent movement of the centrifugal locking bolt. The detents are held back during flight by the friction resulting from spin.

During the period of deceleration in flight the locking bolt and lifting bolt are thrown outwards by centrifugal force, and the graze pellet, assisted by the lifting movement imparted by the step on the lower centrifugal bolt, creeps forward to the limit imposed by the creep wire and thus disengages the recess in the masking bolt. The masking bolt is then thrown clear of the flash-hole leading to the gaine. Further creep of the graze pellet during deceleration in flight is prevented by the creep wire.

On graze, the graze pellet, by its momentum, overcomes the creep wire and moves forward over the needle pellet, carrying the detonator on to the needle. Dependent on the resistance offered by the surface struck and the angle of the strike, the needle pellet may be driven in, severing its shearing wire, and piercing the detonator as the graze pellet moves forward. The flash from the detonator passes through the flash channel to the mixture of lead styphnate and lead azide in the head of the gaine, which brings about the detonation of the cyclonite filling.

Case

The brass case is 7.7 inches long and is not necked. The shell is a close fit in the mouth of the case, and is not secured by coning or indenting.

Propellant Charge

The charge consists of approximately 6 oz. 8 dr. of double base flake propellant in four 46 gram bags. The dimensions of the graphited flakes are .1182 x .1182 x .0138 inches and the composition, nitrocellulose 62.36 per cent. (nitrogen content 12.41 per cent.), nitroglycerine 33.9 per cent., centralite 1.98 per cent., mineral jelly .95 per cent., and graphite .81 per cent.

Primer

The percussion primer used, model 35, is described in this pamphlet.

ITALIAN 47/32 CARTRIDGE, Q.F., A.P. WITH TRACER FUZE
(Fig. 12)

The fixed Q.F. round is fired from the 47 mm. 32 calibre anti-tank gun, model 35.

The complete round weighs 4.5 lb., and has an overall length of 11.8 inches. The shell is pale blue in colour, with a red head, a white band near the centre and a green band immediately above the driving band. The T.N.T. bursting charge is indicated by the stencilling " TRITOLO."

The round consists of the following components:—

> Armour piercing shell filled T.N.T.
> Base fuze with tracer and gaine.
> Brass cartridge case.
> Propellant charge.
> Percussion primer, model 35.

Shell

The shell is approximately 5.4 inches in length and weighs 3.2 lb. filled and fuzed. The body is machined from rolled bar steel and is hardened from the point to a position just above the copper driving band. The V.D.H. hardness figure varies from approximately 690 to 270. The shell is closed at the base by the fuze, which is inserted with a lead washer in front of the flange to act as a gascheck.

The weight of the empty shell is 2 lb. 10½ oz.

Bursting Charge

The bursting charge consists of two pellets of pressed T.N.T. The

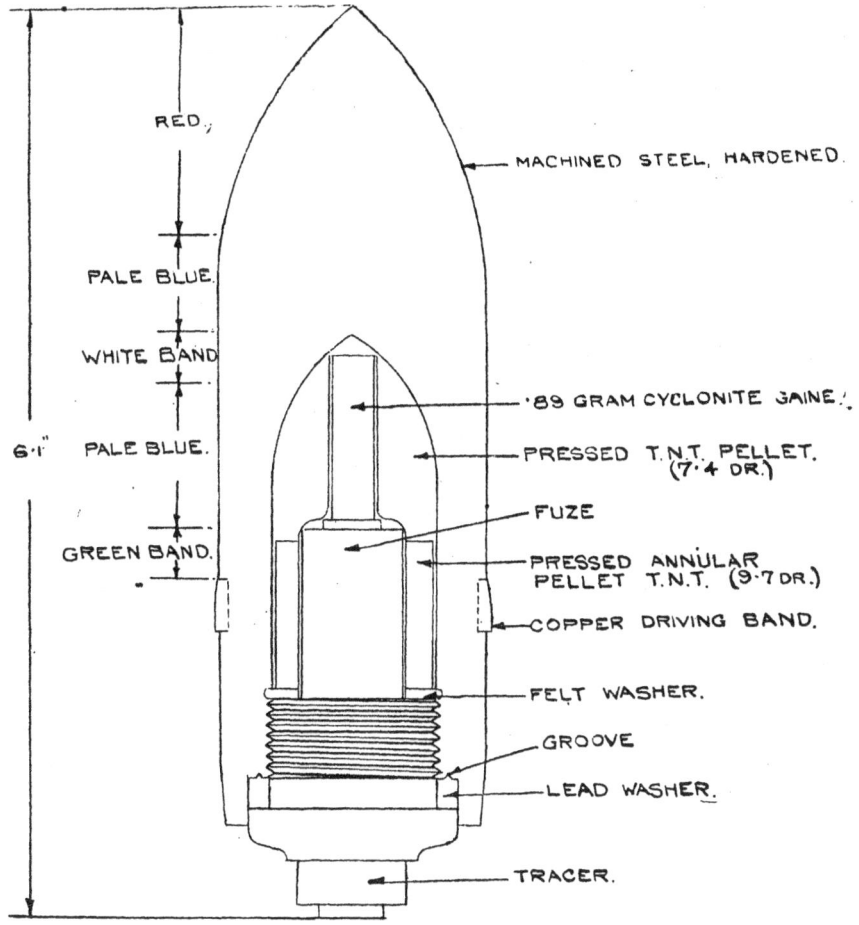

FIG. 12

Italian 47/32 Q.F., Anti-Tank Gun, Mod. 35, A.P. Shell with Tracer Fuze

forward pellet is shaped to fit the ogival head of the shell cavity and has a central cavity to receive the gaine fitted to the fuze. The second pellet is in the form of a hollow cylinder which fits round the forward part of the fuze. A felt washer is inserted between the base of this pellet and the fuze body. The weight of the bursting charge is 1 oz. 1 dr.

Fuze and Tracer (Fig. 13)

The fuze is of the graze type with an igniferous detonator, and is fitted with a .89 gram cyclonite gaine. A tracer filling is contained in the body at the base.

The fuze body is of steel, screw-threaded and flanged externally for insertion in the base of the shell. The diameter of the threaded portion

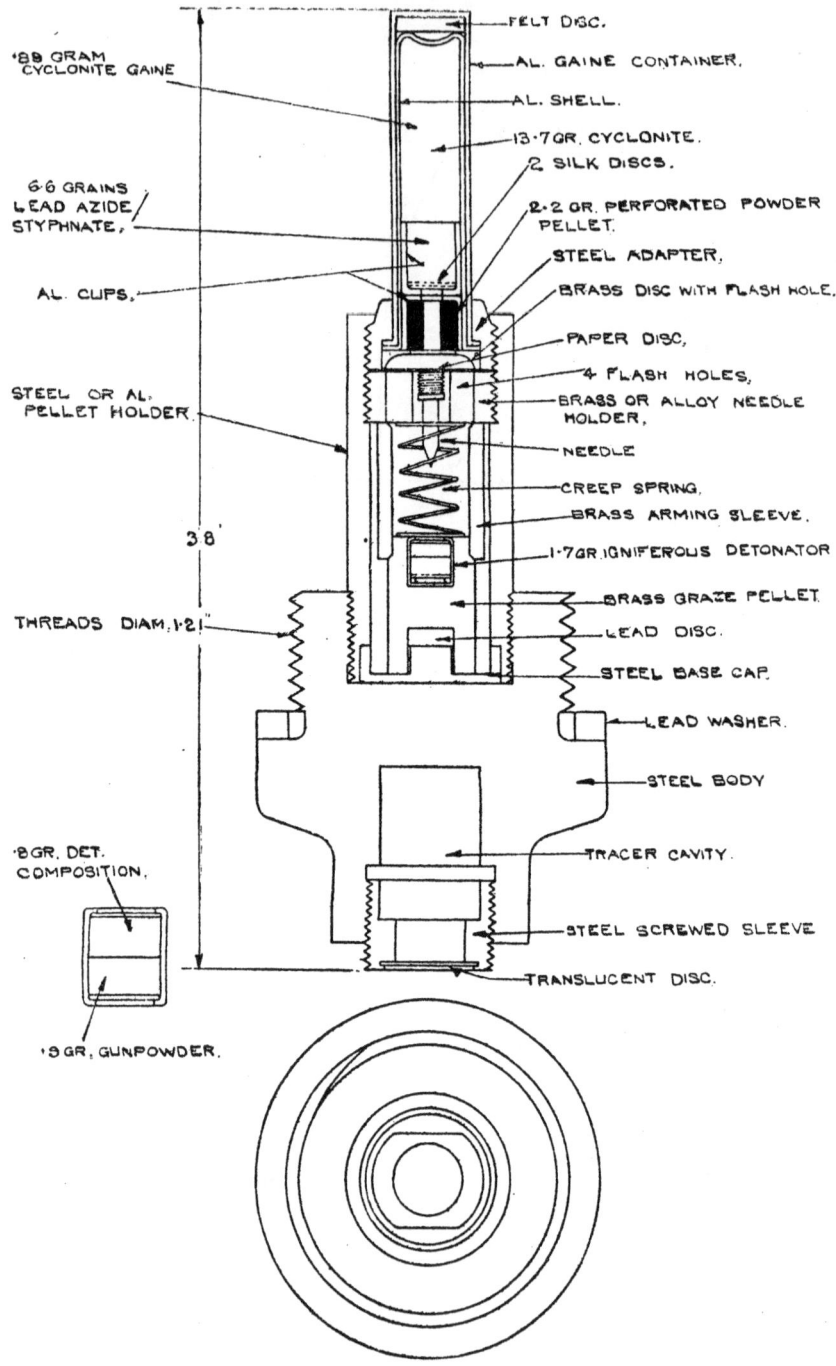

Fig. 13
Italian Base Fuze with Tracer

is 1.21 inches and the pitch approximately 2 mm. A lead washer is fitted in front of the flange to form a gascheck joint. The front face of the body is recessed and screw-threaded to receive the pellet holder.

The pellet holder may be of steel or aluminium, and consists of a hollow cylinder screw-threaded externally at the base end for insertion in the body and internally at the front end to receive the needle holder and gaine adapter. The pellet holder is closed at the base by an internal base cap of steel, which has a central projection carrying a lead disc. The projection and disc fit into a recess in the base of the graze pellet and thus locate the pellet.

The cylindrical brass graze pellet is fitted with an igniferous detonator, and is held off the fixed needle by a creep spring and an arming sleeve. The detonator contains .9 grain of gunpowder primed with a .8 grain pressing of a flash producing detonating composition. The composition consists of:—Mercury fulminate 10.4 per cent., potassium chlorate 48.3 per cent., antimony sulphide 40.2 per cent., and grit 1.1 per cent. A slight shoulder is formed near the front end of the pellet to support the arming sleeve.

The brass arming sleeve is held between the needle holder and the shoulder on the graze pellet and thus prevents the pellet moving towards the needle. The sleeve is split throughout its length to permit expansion during set-back.

The needle holder is screwed into the pellet holder above the arming sleeve, and consists of a screwed steel plug with four flash-holes and a central hole, into which the needle is screwed. The flash-holes are closed by a paper disc.

The .89 gram cyclonite gaine fitted to the front end of the pellet holder is described in this pamphlet in the section on Italian gaines. In this instance the outer cup contains a 2.2 grain perforated pellet of gunpowder.

The tracer, which is fitted in the base of the fuze, and retained by a screwed steel sleeve with a translucent closing disc, consists of a brass cylinder which is open at both ends, and contains the tracing and priming compositions in the form of a pressed pellet. The tracing composition, weighing 32.4 grains, is pressed with drift which leaves a small conical cavity and consists of barium nitrate 61.5 per cent., magnesium 32.6 per cent., with boiled oil and zinc. The priming composition weighs 7.7 grains and is pressed in two increments. The first pressing is made with a serrated drift and the second with a flat drift. The composition consists of barium peroxide 74.1 per cent., magnesium 20.4 per cent., and includes barium carbonate.

Action

On acceleration, the split arming sleeve sets back over the graze pellet and leaves the pellet held off the needle by the creep spring. On graze the pellet overcomes the resistance of the spring by its momentum and impinges the detonator on the needle. The flash produced by the detonator passes through the flash-holes in the needle holder and the brass disc and ignites the powder pellet in the gaine. The flash from the powder pellet initiates the lead azide/lead styphnate mixture, which in turn detonates the cyclonite filling.

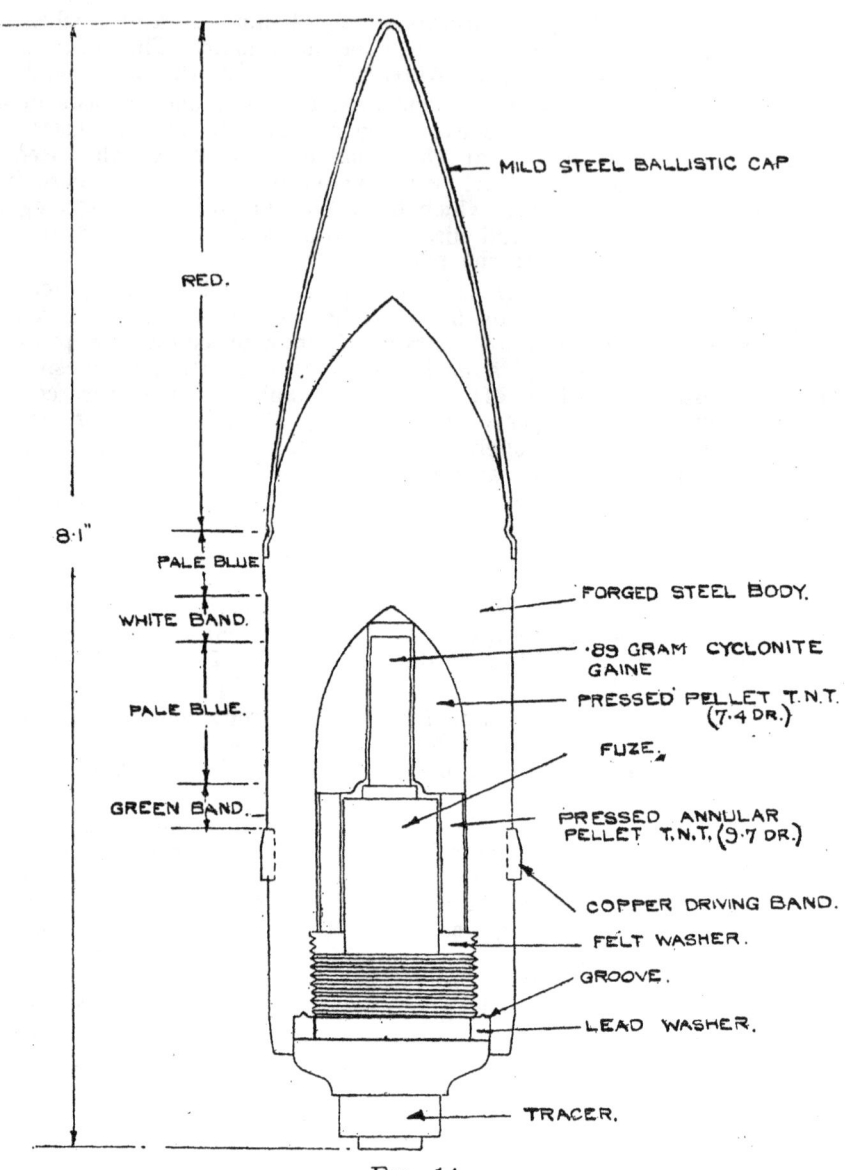

Fig. 14

Italian 47/32 Q.F., A/T Gun, Mod. 35, A.P.B.C. Shell with Tracer Fuze

Case, Propellant Charge and Primer

The brass case is the same as that described for the H.E. round. The propellant charge has a weight of 6 oz. 4 dr. Details of the composition are not yet available. The percussion primer, model 35, is described in this pamphlet.

ITALIAN 47/32 CARTRIDGE, Q.F., A.P.B.C. WITH TRACER FUZE
(Fig. 14)

The fixed Q.F. round is fired from the 47 mm. 32 calibre anti-tank gun, model 35.

The complete round weighs 4.6 lb. and has an overall length of 14 inches. The shell is pale blue in colour, with a red ballistic cap, a white band near the centre and a green band immediately above the driving band. The T.N.T. bursting charge is indicated by the stencilling "TRITOLO."

The round consists of the following components:—
Armour piercing shell with ballistic cap and filling of T.N.T.
Base fuze with tracer and gaine.
Brass cartridge case.
Propellant charge.
Percussion primer, model 35.

Shell
The shell, with the ballistic cap, is approximately 7.4 inches in length, and weighs approximately 3.3 lb. filled and fuzed. The body is 5.4 inches in length, and is forged from bar steel. A cannelure is formed in the head, between two rings of milling, for the attachment of the light mild steel ballistic cap, the cap being pressed into the cannelure. The head of the shell is hardened, the V.D.H. hardness figure varying from approximately 620 at the point to 400 at the shoulder. The shell is closed at the base by the fuze, which is inserted with a lead washer in front of the flange to make a gascheck seal.

The weight of the empty shell with its ballistic cap is 2 lb. $12\frac{1}{4}$ oz.

Bursting Charge
The bursting charge consists of two pressed pellets of T.N.T. similar to those described for the A.P. shell. The weight of the bursting charge is 1 oz. 1 dr.

Fuze, Case, Propellant Charge and Primer
The fuze is the same as that used in the A.P. shell.

The case is the same as that used with the H.E. round. The propellant charge has a weight of 5 oz. 15 dr. Details of the composition are not yet available. The percussion primer, model 35, is described in this pamphlet.

ITALIAN 65/17 CARTRIDGE, Q.F., H.E. WITH FUZE (Fig. 15)

The fixed Q.F. round is fired from the 65 mm. 17 calibre infantry gun.

The complete round weighs 11 lb. and has an overall length of 15.4 inches. The shell is pale blue in colour, with a red head and a green band immediately in front of the driving band. The T.N.T. bursting charge is indicated by the stencilling "TRITOLO."

The round consists of the following components:—
H.E. shell filled T.N.T., with T.N.T. exploder.
Graze fuze with gaine.
Brass cartridge case.
Propellant charge.
Percussion primer, model 35.

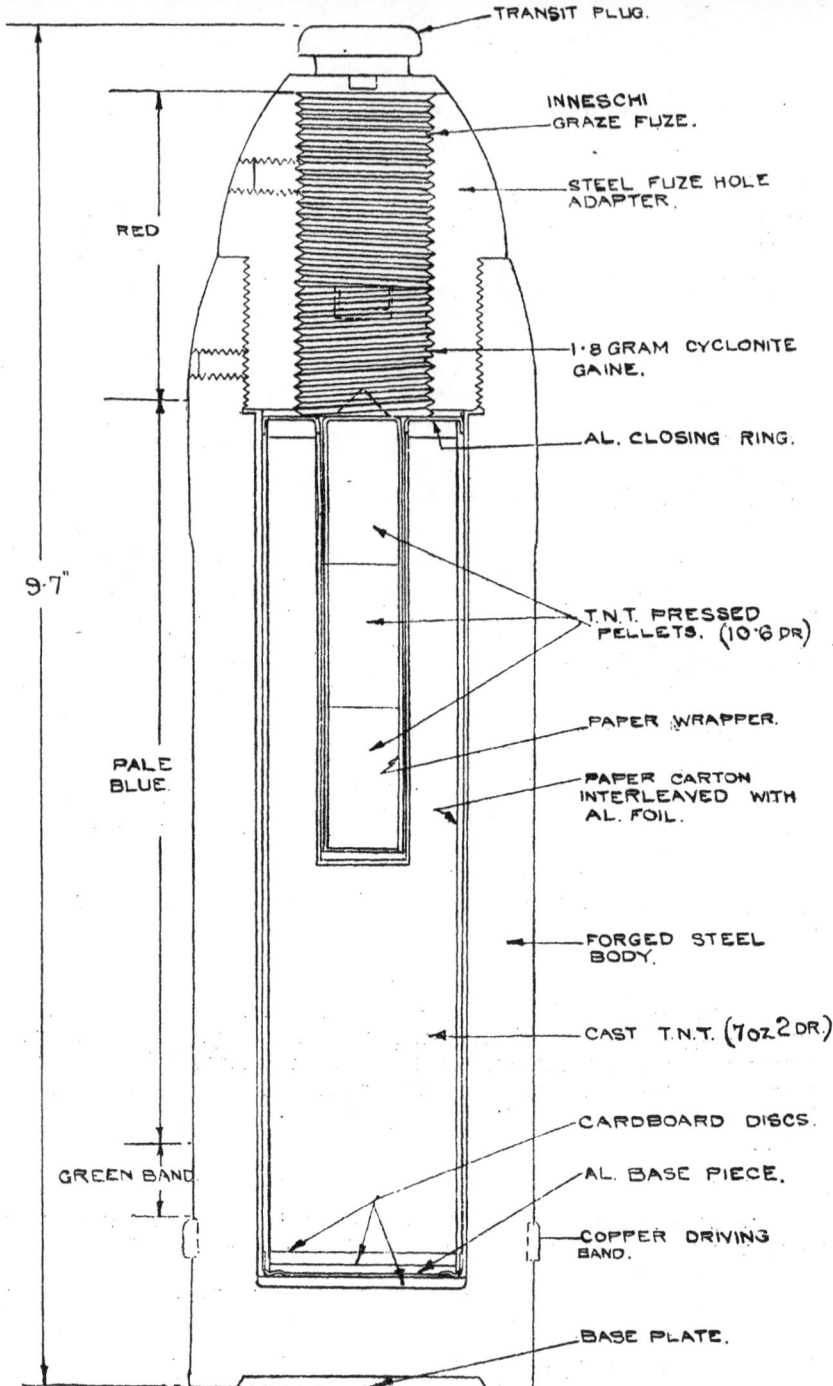

Fig. 15
Italian 65/17 Q.F., Infantry Gun, H.E. Shell

Shell

The shell is of forged steel, and without the fuze-hole adapter is 8 inches in length. Including the adapter, the length is 9.2 inches. The weight filled and fuzed is 9.25 lb. The cavity for the bursting charge is of the parallel walled type, and the shell is fitted with a pressed-in base plate. The driving band is wholly of copper.

The weight of the empty shell is 7 lb. 3 oz., and with the adapter fitted is 8 lb. 4 oz.

Bursting Charge

The bursting charge is of the block type with an exploder cavity, and consists of cast T.N.T. in a carton of paper interleaved with aluminium foil. The carton is closed at the base by two cardboard discs and an aluminium cap, and at the head by a ring of the same material. A cardboard adjusting disc is inserted in the shell cavity beneath the carton. The weight of the bursting charge is 7 oz. 2 dr.

Exploder

The exploder consists of three pressed pellets of T.N.T. in a transparent paper wrapper. The exploder is marked "Alto" near the upper end and "Basso" near the lower end. The density of the top pellet is 1.45 and of the other two 1.5. The weight of the exploder is 10.6 dr.

Fuze and Gaine

The Inneschi graze fuze and the 1.8 gram cyclonite gaine used in this shell are described in this pamphlet.

Case

The brass cartridge case is 6.7 inches long and is not necked. The shell is a close fit in the mouth of the case and is not secured by coning or indenting.

Propellant Charge

The charge consists of 5 oz. 10 dr. of double base propellant flake contained in a bag. A folded strip of tinfoil is placed on top of the charge to act as a decoppering agent, and the charge is held in position over the primer by a cardboard cup surmounted by a cardboard distance piece, above which is a second cardboard cup and dried vegetable packing.

Primer

The percussion primer, model 35, which is used is described in this pamphlet.

ITALIAN 75/27 Q.F. CARTRIDGE

The cartridge is of the separate loading Q.F. type and is used in the 75 mm. 27 calibre gun-howitzer.

The brass case is fitted with a model 29 percussion primer and contains a 14 oz. 13 dr. propellant charge in four sections and a decoppering agent in the form of a crumpled piece of foil. The mouth of the

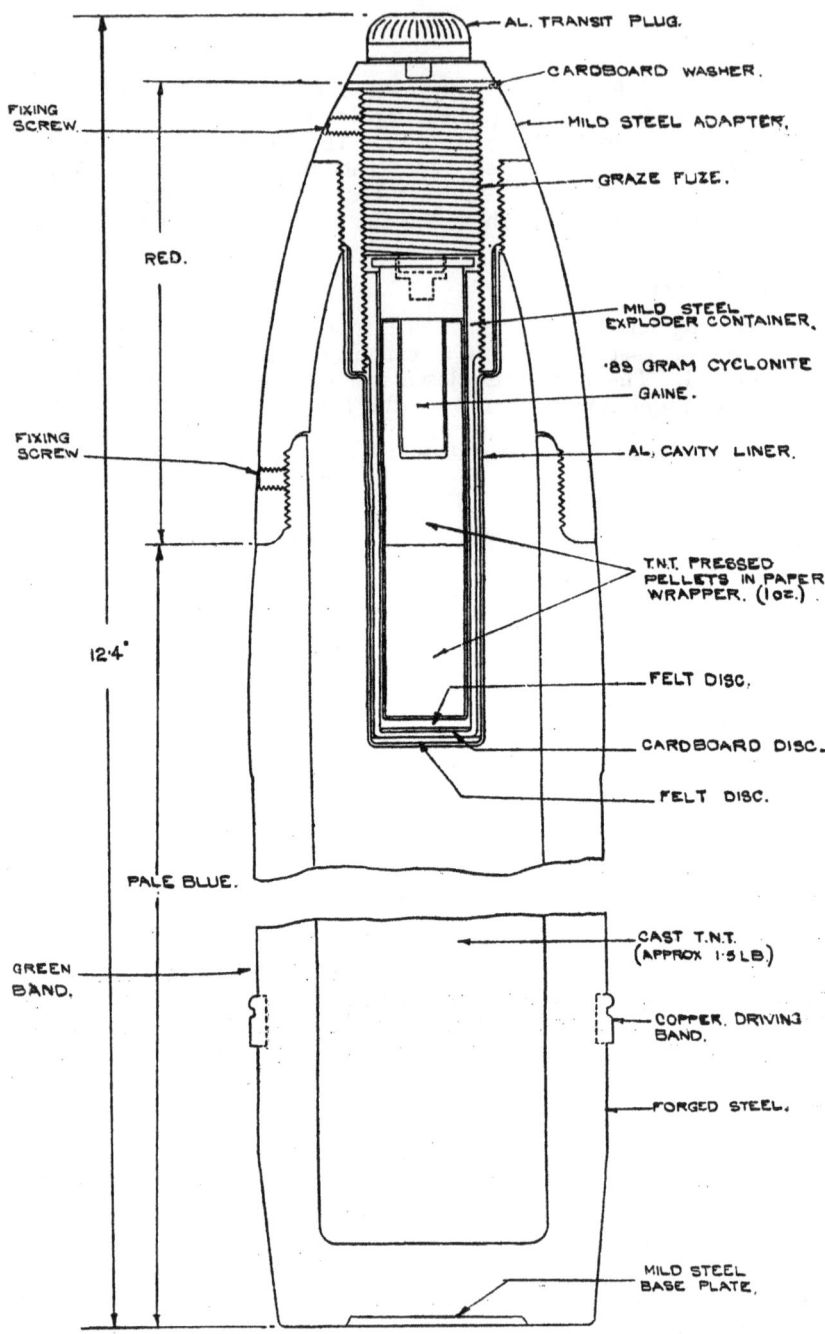

FIG. 16
Italian 75/27 Q.F., Gun-How., Streamlined H.E. Shell, Mod. 32

case is closed by a cardboard cup. The case is approximately 7.4 inches in length and is stamped " 75/27 " on the base.

The propellant charge consists of four bags, one of 240 grams and three of 60 grams of flake propellant. The dimensions of the flake are .47 × .47 × .047 inches. Details of the composition of the propellant are not yet available. The 240 gram bag is " Charge 1 " and is marked " Elemento Fondamentale." The 60 gram bags are each marked " Elemento Aggiuntivo." Each of the bags is also marked to indicate the weight of the section, the size of the flake, and with the designation " 75/27."

ITALIAN 75/27, Q.F., GUN-HOW., STREAMLINED, H.E. SHELL, MOD. 32 (Fig. 16)

The shell is fired from the Q.F. 75 mm. 27 calibre gun-howitzer with a separate loading Q.F. cartridge.

The head of the shell, including the fuze-hole adapter, is red down to its junction with the body. The remainder of the shell is pale blue, with a green band above the driving band, and is stencilled " 75/27." The T.N.T. bursting charge is indicated by the stencilling " TRITOLO." The overall length of the fuzed shell is 12.4 inches, and the weight filled and fuzed is 13 lb. 14 oz. 6 dr.

Shell

The streamlined forged steel shell is in two main parts, the ogival head being screwed on above the shoulder and secured by a fixing screw. Circular recesses are formed in the head to receive the assembly tool. The shell is fitted with a copper driving band with a single cannelure, and has a comparatively thin wall of parallel section. A mild steel base plate is pressed into the base.

The fuze-hole adapter is of mild steel and is shaped at the head to correspond with the contour of the shell. The adapter is screw-threaded externally for insertion in the shell and internally to receive the fuze and the exploder container. A fixing screw is provided for the fuze, and circular recesses are formed to receive the assembly tool.

The exploder container is of mild steel, and consists of a tubular pocket about 3.8 inches in depth, with an external screw-thread at its upper end for insertion in the adapter.

The weight of the empty shell is 12 lb. 1 oz. 10 dr.

Bursting Charge

The bursting charge consists of approximately 1 lb. 6 oz. 7 dr. of cast T.N.T., with a 5.3 inch cavity to accommodate the exploder container and the lower part of the adapter. The upper part of the cavity is enlarged in diameter to receive the adapter, and the whole of the cavity wall is supported by an aluminium liner. A felt disc is inserted between the base of the liner and the base of the exploder container.

Exploder

The exploder consists of two cylindrical pressed pellets of T.N.T. enclosed in a transparent paper wrapping. The upper pellet has a cavity to accommodate the gaine. The wrapper is marked " Per.

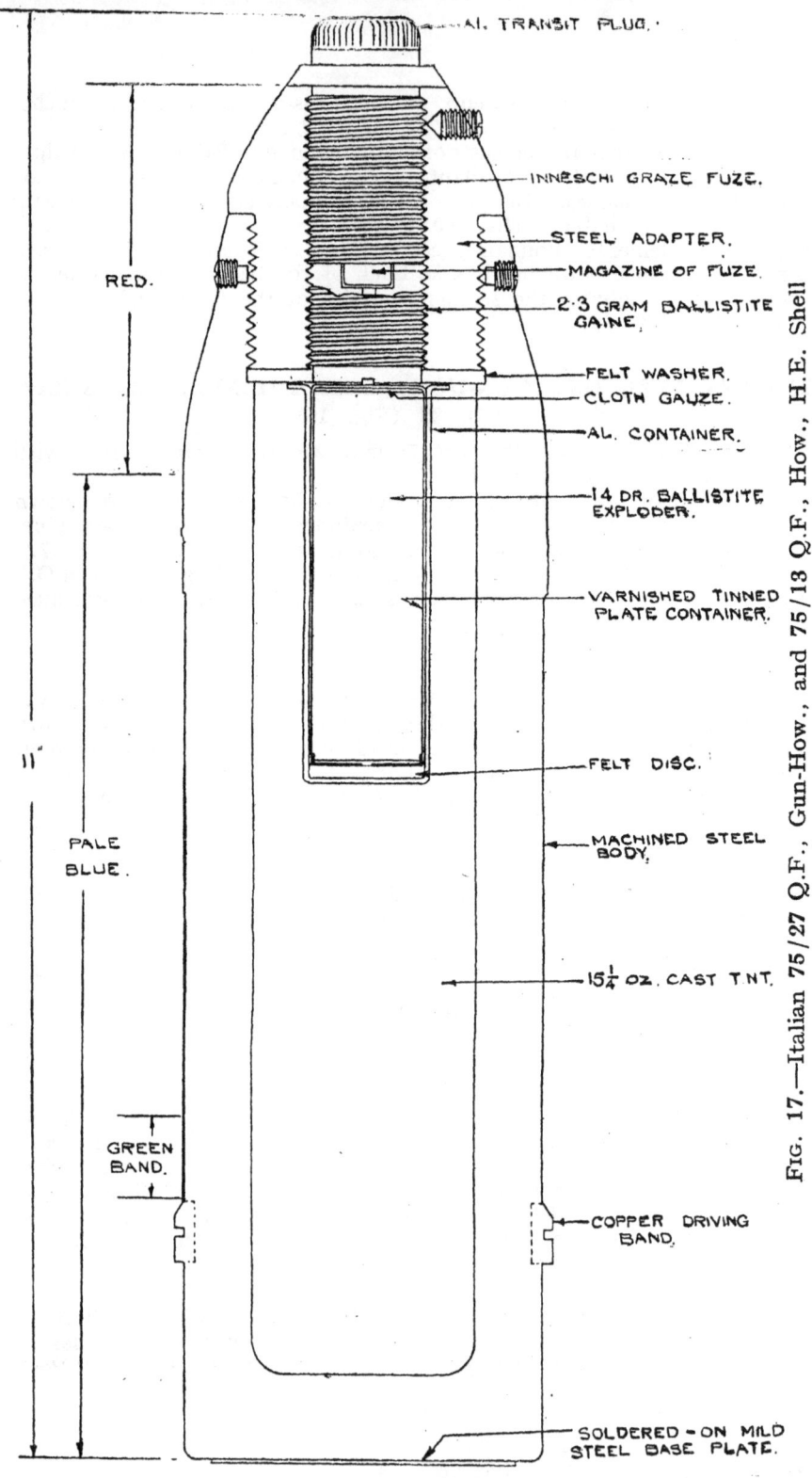

Fig. 17.—Italian 75/27 Q.F., Gun-How., and 75/13 Q.F., How., H.E. Shell

Deton. A.E. No. 1." A cardboard disc, followed by a felt disc, is inserted in the exploder container beneath the exploder. The weight of the exploder is 1 oz.

Fuze and Gaine

A description of the fuze, the Inneschi graze fuze, and the .89 gram cyclonite gaine used in this shell are included in this pamphlet.

ITALIAN 75/13 Q.F. CARTRIDGE

The cartridge is of the separate loading Q.F. type and is used in the 75 mm. 13 calibre howitzer.

The brass case is fitted with a Model 29 percussion primer and contains an 8 oz. 7 dr. propellant charge in four sections. The mouth of the case is closed by a cardboard cup. The case is 5.18 inches in length and is stamped on the base " 75 cm. M.15."

The propellant charge consists of four bags, one of 120 grams and three of 40 grams of flake propellant. The dimensions of the flake are .394 x .394 x .039 inches. Details of the composition of the propellant are not yet available. The 120 gram section is " Charge 1 " and is marked " Elemento Fondamentale." The 40 gram bags are each marked " Elemento Aggiuntivo." Each of the bags is also marked to indicate the weight of the section, the size of the flake, and with the designation " 75/13."

ITALIAN 75/27, Q.F., GUN-HOW. AND 75/13, Q.F., HOW. H.E. SHELL (Fig. 17)

The shell is fired from the Q.F. 75 mm. 27 calibre gun-howitzer and the Q.F. 75 mm. 13 calibre howitzer, with a separate loading Q.F. cartridge.

The head of the shell, including the fuze-hole adapter, is red. The remainder of the shell is pale blue, with a green band above the driving band, and is stencilled " 75/13/27." The T.N.T. bursting charge is indicated by the stencilling " Tritolo." The overall length of the fuzed shell is 11 inches and the weight, filled and fuzed, is 13 lb. 11 oz. 8 dr.

Shell

The shell is machined from rolled bar steel and has a cylindrical band or neck formed in the head, with two fixing screws for the fuze-hole adapter. The driving band is of copper, and a mild steel base plate is attached by soldering. The cavity for the bursting charge is of the parallel wall type.

The fuze-hole adapter is of steel and is shaped at the head to form the nose of the assembled projectile. The adapter is screw-threaded externally for insertion in the shell and internally to receive the fuze and the gaine. A fixing screw is provided for the fuze.

The weight of the empty shell, including the adapter, is 12.1 lb. The adapter alone weighs 1.18 lb. The length of the shell body is 9.5 inches. With the adapter fitted, the length is 10.5 inches.

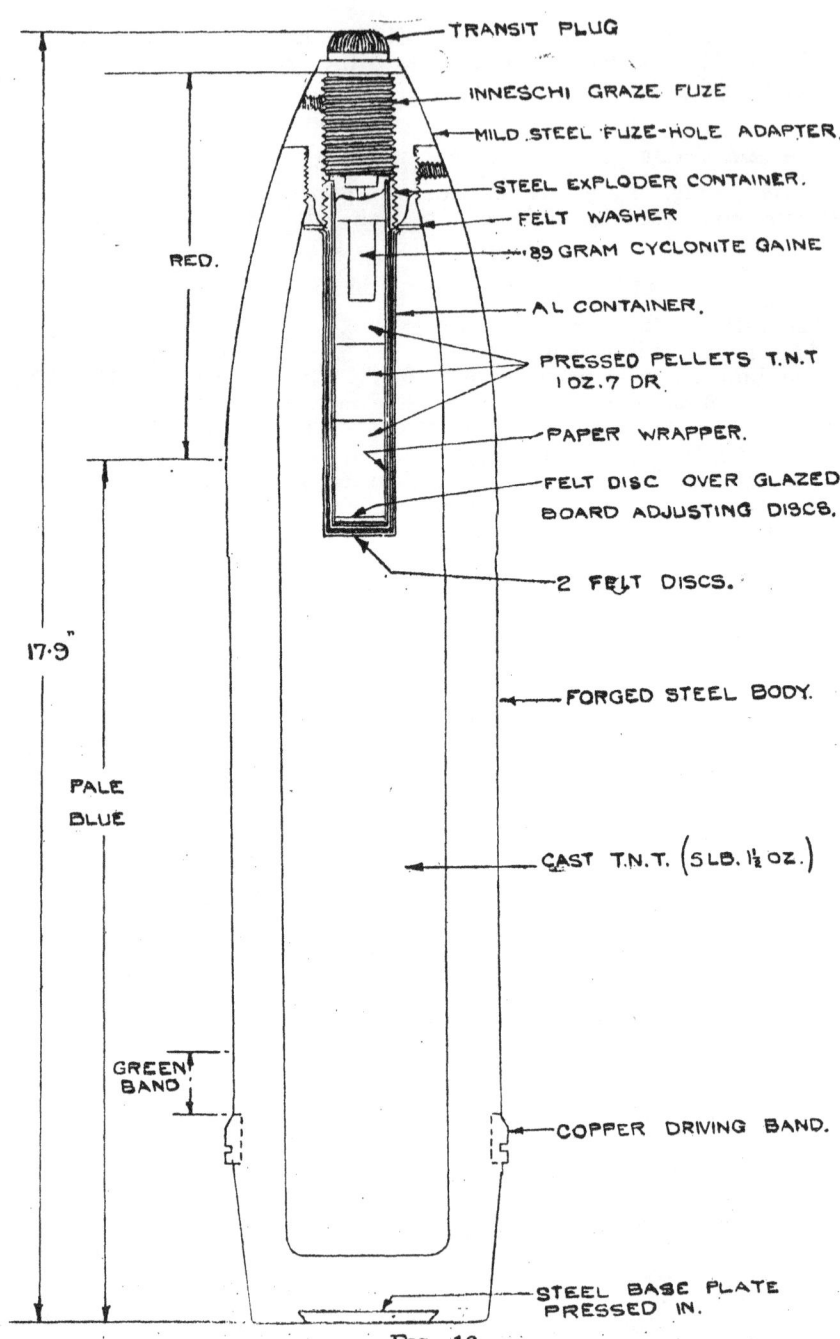

Fig. 18
Italian 100/17 Q.F., How., H.E. Shell

Bursting Charge

The bursting charge consists of cast T.N.T., with an exploder cavity, the wall of which is supported by an aluminium exploder container. A felt washer is inserted between the top of the bursting charge and the underside of the adapter. The weight of the bursting charge is 15.25 oz.

Exploder

The exploder consists of 14 drams of ballistite flake in a cylindrical container of varnished tinned plate. The container is 2.9 inches in length and .87 inch in diameter, with a small central hole at each end. The holes are lightly closed by a cloth gauze disc on the inner face of the disc forming the end. The ballistite consists of 51 per cent. of nitrocellulose (nitrogen content 13 per cent.) and 49 per cent. of nitroglycerine. Chalk is included as a stabilizer.

Fuze and Gaine

A description of the fuze, the Inneschi graze fuze, and the 2.3 gram ballistite gaine used in this shell are included in this pamphlet.

ITALIAN 100/17 Q.F. CARTRIDGE

The cartridge is of the separate loading type, and is used in the 100 mm. 17 calibre howitzer. The brass case is fitted with a Model 35 percussion primer, and contains a 1 lb. 3 oz. 12 dr. propellant charge in five sections. The mouth of the case is closed by a cardboard cup. The case is 5.2 inches in length.

The propellant charge consists of five bags, one of 280 grams and four of 70 grams of flake propellant. The dimensions of the flake are .59 x .59 x .059 inches. Details of the composition are not yet available. The 280 gram section is "Charge 1" and is marked "Elemento Fondamentale." The 70 gram sections are marked "Elemento Aggiuntivo." Each of the sections is marked with the designation of the equipment, the weight of the section and the dimensions of the flake.

ITALIAN 100/17, Q.F., HOWITZER H.E. STREAMLINED SHELL
(Fig. 18)

The shell is fired from the Q.F. 100 mm. 17 calibre howitzer, with a separate loading cartridge.

The head of the shell, including the fuze-hole adapter, is red. The remainder of the shell is pale blue, with a green band above the driving band, and is stencilled 100/17. The T.N.T. bursting charge is indicated by the stencilling "TRITOLO." The overall length of the fuzed shell is 17.9 inches, and the weight, filled and fuzed, is 29 lb. 11 oz.

Shell

The shell is of forged steel, with a copper driving band, and is streamlined from immediately behind the driving band. A steel base plate is pressed into a recess in the base. The cavity for the bursting charge is of the parallel walled type, but, following the external contour of the shell, the cavity decreases in diameter towards the nose. A fixing screw is fitted at the nose to secure the fuze-hole adapter.

The adapter is of mild steel and is shaped at the head to coincide with the contour of the shell. The adapter is screw-threaded externally

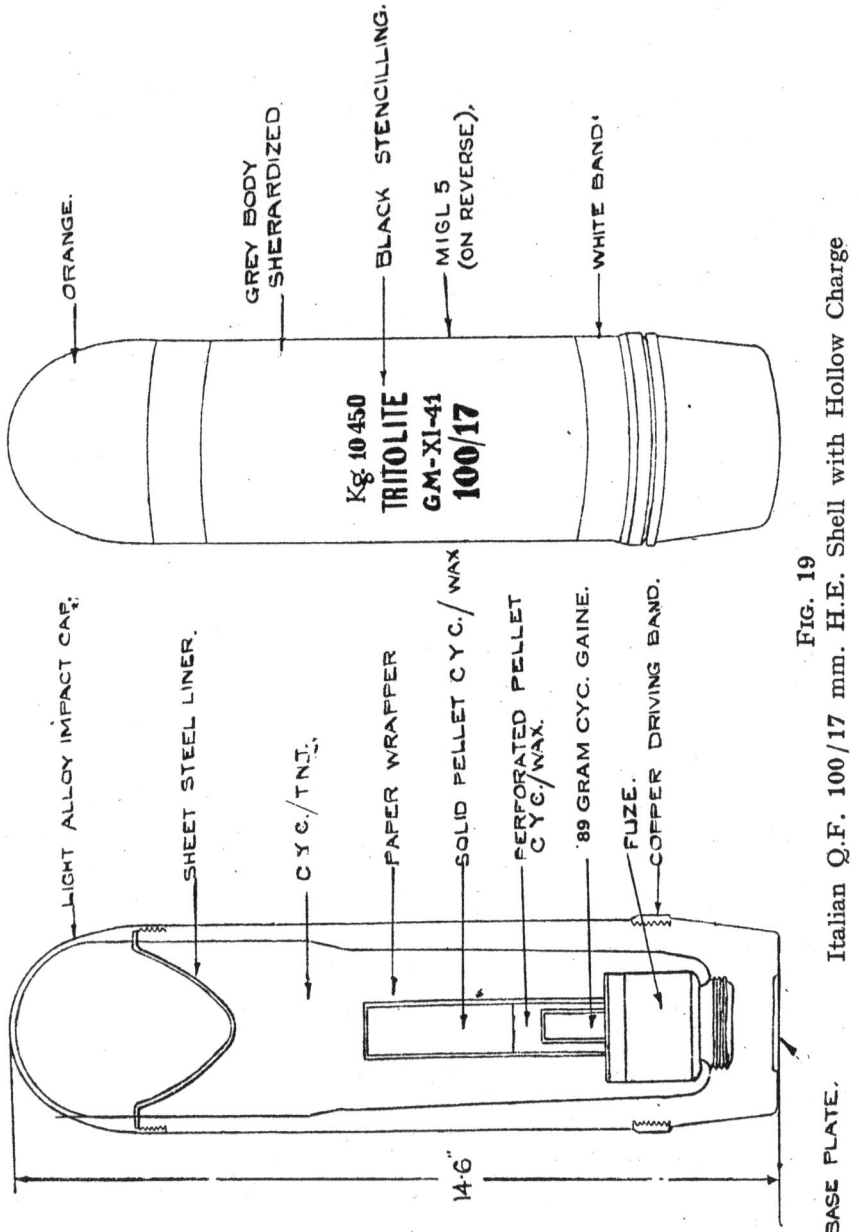

Fig. 19

Italian Q.F. 100/17 mm. H.E. Shell with Hollow Charge

for insertion in the shell and internally to receive the fuze and the steel exploder container. A fixing screw is provided for the fuze.

The weight of the empty shell, including the adapter, is 24.3 lb. The adapter weighs .76 lb. The length of the shell body is 16.3 inches. With the adapter fitted, the length is 17.4 inches.

Bursting Charge

The bursting charge consists of cast T.N.T. with an exploder cavity the wall of which is supported by an aluminium container. A felt washer is inserted between the top of the bursting charge and the underside of the adapter. Two felt discs are inserted in the bottom of the aluminium container. The weight of the bursting charge is 5 lb. $1\frac{1}{2}$ oz.

Exploder

The exploder consists of three pressed cylindrical pellets of T.N.T. in a transparent paper wrapper. The upper pellet has a cavity to receive the gaine. The length of the exploder is 4.1 inches and the diameter .73 inch. The weight is 1 oz. 7 dr. Adjusting discs of glazed board are used in the bottom of the steel exploder container under a felt disc to ensure a good contact between the gaine and the exploder.

Fuze and Gaine

A description of the Inneschi graze fuze and the .89 gram cyclonite gaine used in the shell is included in this pamphlet.

ITALIAN 100/17, Q.F., HOW. HOLLOW CHARGE SHELL WITH INTERNAL BASE FUZE (Fig. 19)

The shell is used in the Q.F. 100 mm. 17 calibre howitzer, with a separate loading Q.F. cartridge, and is designed with a " hollow " at the forward end of the bursting charge for the perforation of armour by blast.

The markings and dimensions of the shell, for identification, are shown in the drawing. The weight of the filled projectile is approximately 23 lb. 2 oz.

The streamlined body of the shell has a comparatively thin wall, and is fitted with a copper driving band and a pressed-in mild steel base plate. An external screw-thread is formed at the head for attachment of the dome-shaped impact cap of light alloy, which brings about the function of the base graze fuze and the detonation of the bursting charge at an effective distance from the target. Internally a screw-threaded recess is formed in the base of the cavity for the insertion of the fuze. The walls of the cavity are treated with paraffin wax. The weight of the empty shell, without fuze, is 18 lb. 10 oz.

An exploder, consisting of two pressed pellets of cyclonite/wax (one solid and the other perforated) wrapped together in a waxed or varnished paper wrapper, is fitted over the paper sleeve covering the gaine fitted to the fuze. The solid pellet weighs 540 grains and contains a small quantity of pink dye. The perforated pellet weighs 332 grains. The paper wrapper is marked " Det. Sec. A.E.—T4F 5 per cent. N 2 MOD 38."

The bursting charge is cast at a density of 1.63 and consists of 3 lb. 9 oz. of cyclonite/T.N.T. (60/40). The cavity or hollow formed in the head of the bursting charge is approximately parabolic in shape, and is supported by a steel liner, which is secured by the impact cap and has a maximum diameter of 2.85 inches and a thickness of approximately .047 inch.

Fuze (Fig. 20)

The fuze is of the detonating type, with a graze action, and is screwed into a recess in the base of the shell cavity.

The cylindrical body contains a screwed aluminium pellet holder, and is closed at the top by a screwed plug which fits over the fixed striker in the pellet holder and carries the gaine. The body is open at the base.

The pellet holder has a central recess which contains the graze or inertia pellet, and is closed at the base by a screwed cap of aluminium. The recess is reduced in diameter near the top to engage the shearing wire fitted in the pellet. An adjacent recess is fitted with a detent and spring in an aluminium sleeve. The detent supports a brass locking pin which locks the centrifugal bolt, between the striker and detonator, in a transverse channel formed in the holder. The channel is closed at each end by an aluminium disc. The centrifugal bolt is a hollow cylinder of aluminium with an aperture to receive the locking pin. Two channels extending through the length of the holder are closed at the top by aluminium discs, and at the base communicate with the recess containing the graze pellet. These, in conjunction with escape holes formed below the detonator in the pellet, are apparently intended to provide an escape for the pressure if the detonator is accidentally initiated. The steel striker, with four flash-holes formed round it, is screwed into a projection on the top of the holder.

The brass graze pellet is held in position in the holder by the centrifugal bolt and is fitted with a protruding shearing wire which does not come into contact with the holder until the shell is in flight. The detonator in the head of the pellet is of the igniferous type, and consists of a copper sleeve with an inner cup of tinfoil which contains .77 grain of a composition consisting of:—Mercury fulminate 11 per cent., potassium chlorate 50.5 per cent., antimony sulphide 38.5 per cent. A recess below the detonator communicates with escape holes leading to the exterior of the pellet.

Gaine

A description of the .89 gram cyclonite gaine fitted to the fuze is included in this pamphlet.

Action

On acceleration the detent and locking pin set back, thereby releasing the centrifugal bolt. During this period the bolt is held by the set back. During flight the bolt is moved clear of the graze pellet and the striker by centrifugal force, and deceleration causes the graze pellet to creep forward until its shearing wire fouls the step at the forward end of the recess in the holder. The igniferous detonator in the pellet is then very near to the fixed striker and is only held clear by the shearing wire. On graze the shearing wire is overcome and the pellet carries the detonator on to the striker. The flash produced passes through the flash-holes round the striker and initiates the lead azide and styphnate filling which brings about the detonation of the cyclonite filling of the gaine.

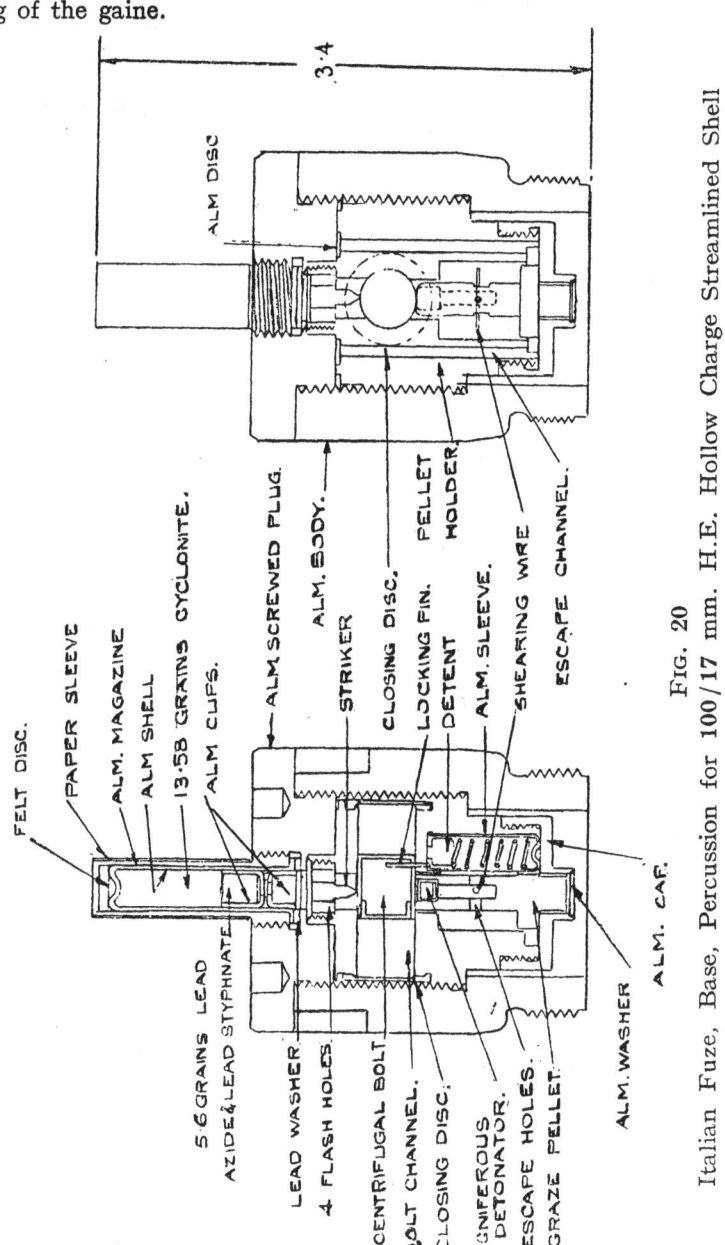

Fig. 20

Italian Fuze, Base, Percussion for 100/17 mm. H.E. Hollow Charge Streamlined Shell

GERMAN PRIMER, PERCUSSION, Q.F. CARTRIDGE, C/13nA
(Fig. 21)

The brass body of the primer is .56 inch long and has a diameter of .52 inches at the screw-threaded portion. Three equi-spaced recesses for the primer key are formed round the circumference of the base. The model number, C/13, followed by the letters " nA " indicating new pattern, is stamped in the base.

The brass cap is positioned inside the base of the body, under the anvil, and is not visible from the exterior. The cap contains a .33 grain filling consisting of mercury fulminate 52 per cent., potassium chlorate 23 per cent., antimony sulphide 19.7 per cent., and grit 5.3 per cent. The composition is sealed by a covering of varnish without the addition of a foil disc.

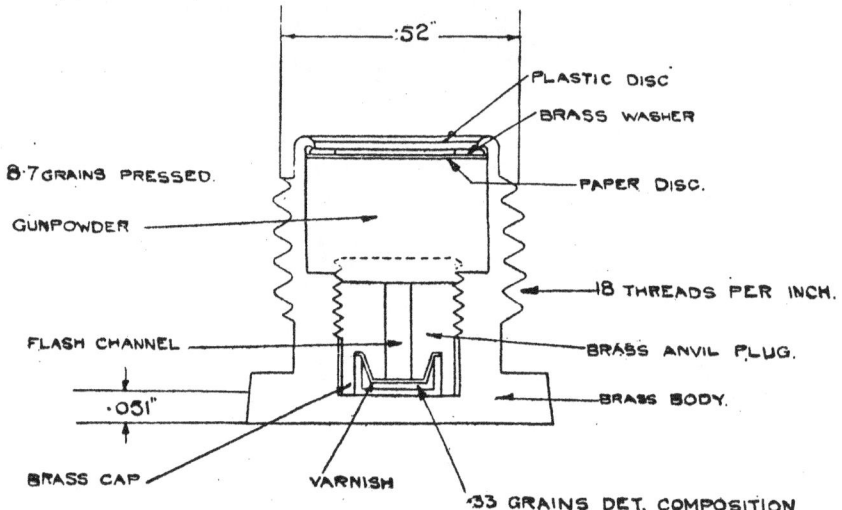

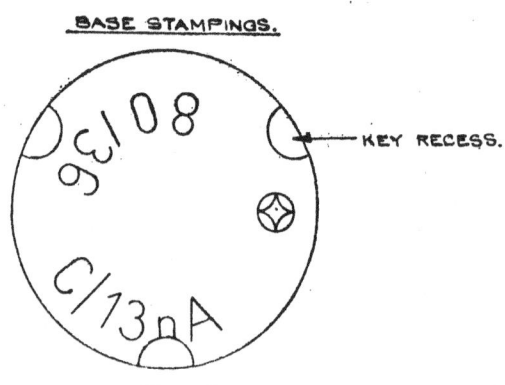

Fig. 21
German Primer, Percussion, Q.F. Cartridge, C/13nA

The brass anvil plug is screwed into the body and is shaped at the base to receive the cap. A flash channel is formed through the centre of the plug.

The magazine formed in the upper part of the body contains 8.7 grains of pressed gunpowder covered by a paper disc. The magazine is closed by a brass washer covered by a plastic disc. The disc is secured by the rim of the body being turned inwards.

GERMAN PRIMER, PERCUSSION, Q.F. CARTRIDGE, C/33
(Fig. 22)

The brass body of the primer is .55 inch long and has a diameter of approximately .5 inch at the screw-threaded portion. Two key flats are formed in the base, which is stamped with the model number C/33.

The brass cap is positioned inside the base of the body, under the anvil, and is not visible from the exterior. The cap contains a .37 grain filling consisting of mercury fulminate 24.6 per cent., potassium chlorate 37.6 per cent., antimony sulphide 29.6 per cent., and ground glass 8.2 per cent. The filling is covered by a disc of tinfoil without varnish.

The anvil plug is of brass and is screwed into the body over the cap. A screw slot is cut in its front face, and two slots are formed, diametrically opposite, down the side to provide flash channels.

The magazine formed in the upper part of the body contains a 2 grain filling of granular gunpowder immediately above the anvil plug, over which is a 6 grain pellet of gunpowder. The pellet is covered by a paper disc and the magazine closed by a plastic disc inserted over a brass washer. The closing disc is held by the turned-in rim of the body.

GERMAN 2 CM. SOLOTHURN S.A. CARTRIDGE WITH A.P./T. SHOT (TUNGSTEN CARBIDE CORE) (2 cm. Pzgr. 40) (Fig. 23)

The complete round has an overall length of 7.36 inches. The details corresponding to those tabulated in Pamphlet No. 5 for other 2 cm. Solothurn cartridges are as follows:—

(a) Body of shot unpainted except below the driving band, where it is lacquered black.
(b) The body is marked with a .2 inch red band above the driving band.
(c) The weight of the tungsten carbide core is 960 grains.
(d) The total weight of the shot is 1,555 grains.
(e) The propellant charge consists of 679 grains of N.C.T.
(f) The colour of the trace is white, changing to red.
(g) The weight of the complete round is 4,300 grains (approx. 9 oz. 13 dr.).

The percussion cap in the base of the case is of steel and the cap annulus is lacquered red.

The shot consists of a duralumin body which has a conical head and contains a tungsten carbide core, and is fitted with a tracer screwed in at the base. The body is increased in diameter at the shoulder, and

Head of Anvil Plug.

BASE STAMPINGS.

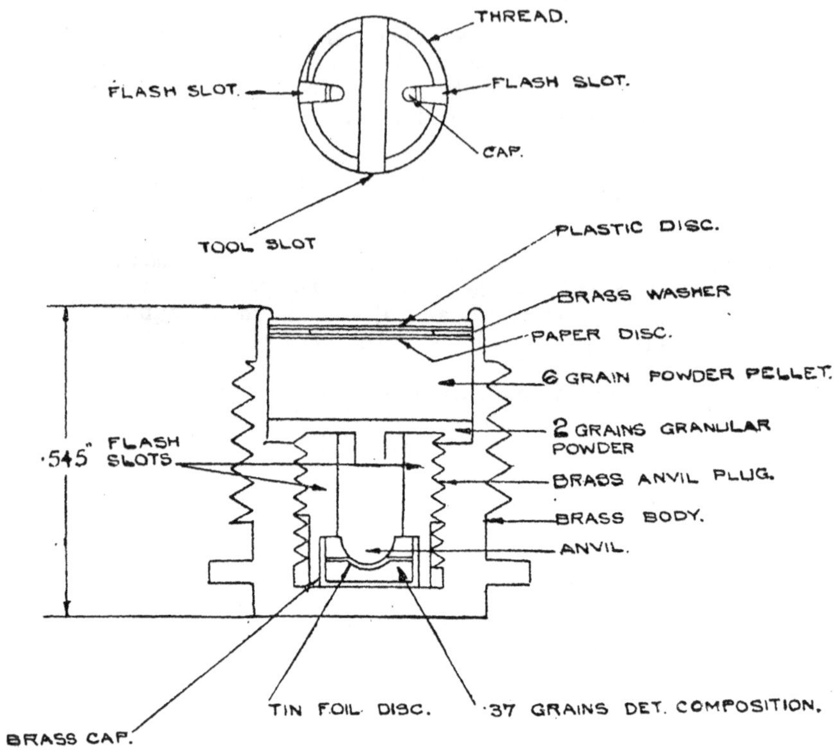

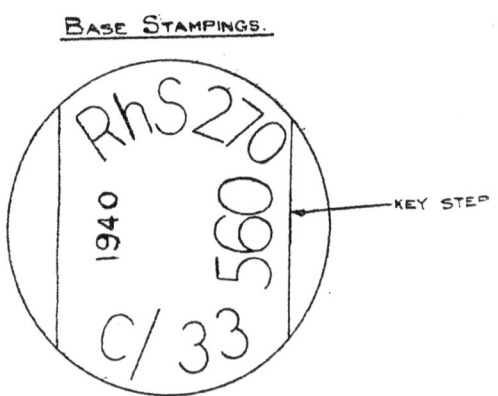

FIG. 22

German Primer, Percussion, Q.F. Cartridge, C/33

is fitted with a copper driving band and cannelured near the base. The armour-piercing core has a V.D. Hardness figure of about 1,700, and the composition, by analysis, is as follows:—

Carbon	4.2	per cent.
Tungsten	92.0	,,
Niobium	2.4	,,
Iron	1.2	,,

Cobalt, nickel, titanium, silicon—trace or nil.

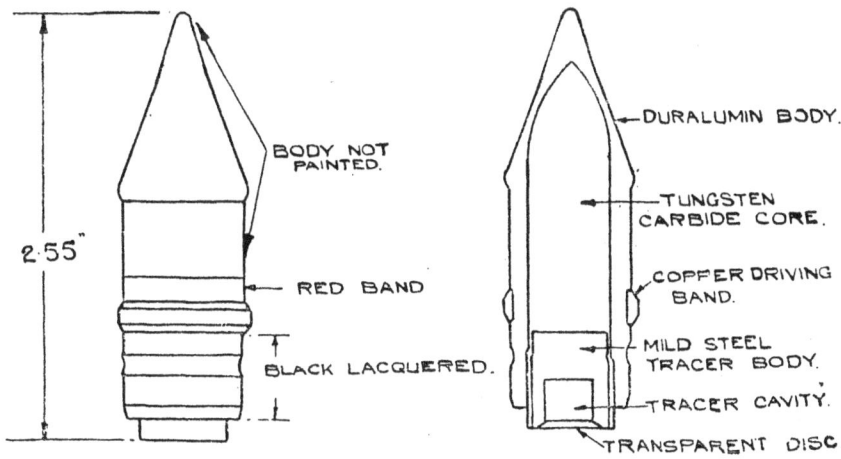

FIG. 23

German 2 cm. Solothurn A.P./T. Shot, with Tungsten Carbide Core

This composition differs from previous types in that an iron bond and not a nickel bond is used. The head of the core is ogival and is struck with a .7 inch radius.

The tracer consists of a mild steel body, which is screwed into the shot behind the A.P. core, and contains the tracing and priming composition in a cavity in its base. The cavity is closed by a transparent disc secured by ringing.

The following stampings have been met with:—

" Aux 1 K 41 WaA109 " on the shot 2 inches above the driving band.
" Avu WaA73 41 6h " on the base of the case.

GERMAN 3.7 CM. PAK CARTRIDGE, Q.F., A.P. WITH TRACER FUZE (3.7 Panzergranat-Patrone) (Fig. 24)

The fixed Q.F. round is used in the 3.7 cm. anti-tank gun, and has an overall length of 13.3 inches. The weight of the complete round is approximately 2 lb. 14 oz. The shell is painted black.

The complete round consists of the following components:—

Armour-piercing shell filled P.E.T.N./Wax.
Base fuze with tracer, model 5103 or 5103* and gaine.
Case model 6331.

Propellant charge of the double base type in tubular form with decoppering **agent** and nitrocellulose igniter.
Primer percussion, model C/13.

Shell

The shell is machined from rolled bar steel and is hardened. The V.D. Hardness figures vary from approximately 600 at the head to 500

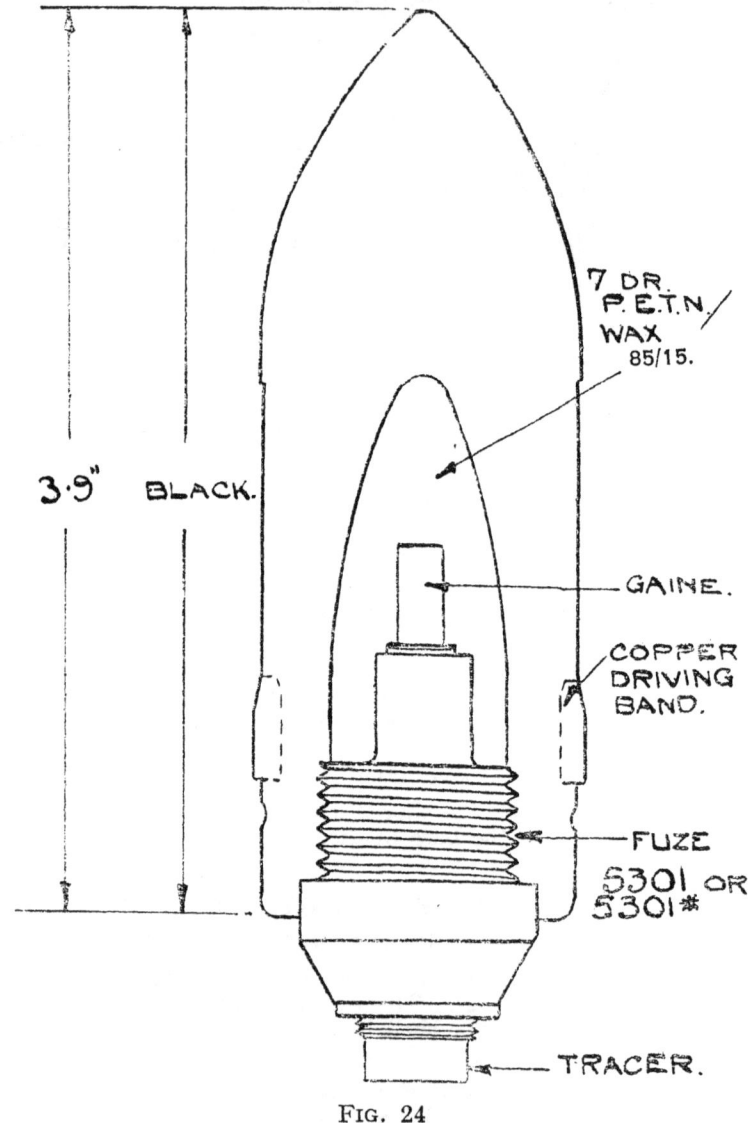

Fig. 24
German 3.7 cm. Pak A.P. Shell

at the base. In section the shell has the solid head and small cavity characteristic of piercing shell. The weight of the shell, filled and fuzed, is approximately 1.5 lb. The driving band is wholly of copper, and behind it the shell is cannelured for the attachment of the case. The shell is closed at the base by the fuze. The weight of the empty shell is 1 lb. 4½ oz.

Bursting Charge

The bursting charge consists of approximately 7 drams of P.E.T.N./Wax (85/15), with a two diameter cavity for the gaine and fuze.

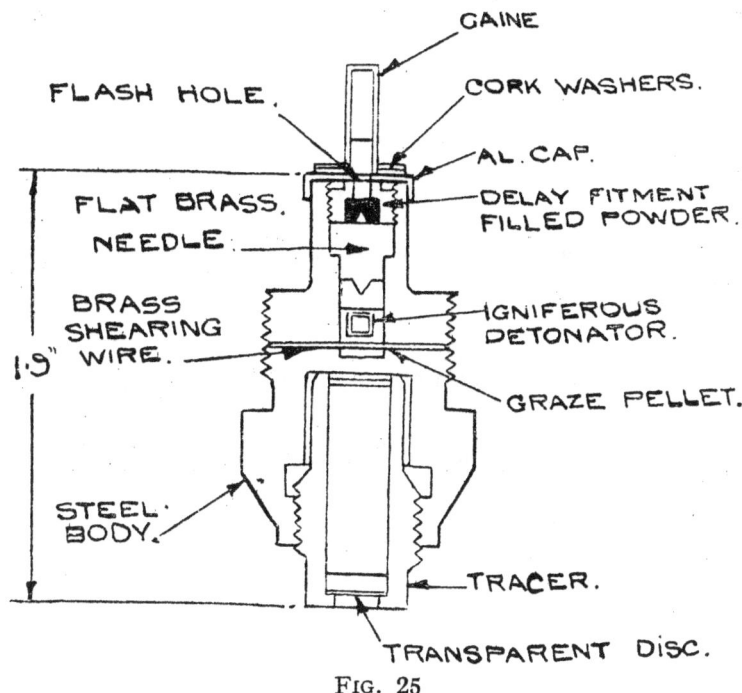

Fig. 25

German Base Fuze 5103 and Gaine

Fuze

Either the fuze "Bd.Z.5103" or "Bd.Z.5103*" may be used. A description and drawing of fuze 5103* are included in Pamphlet No. 4 on pages 14 and 15. The drawing, Fig. 8, shows the fuze fitted with an adapter and gaine which are not used in the 3.7 cm. shell. Instead the fuze is closed at the front end by an aluminium cap with a central flash-hole, and a gaine is situated above it in the shell. These details are the same as those shown in the drawing of the fuze 5103 (Fig. 25).

The fuze 5103 (Fig. 25) differs from the 5103* in having a delay fitment with a powder pellet instead of being fitted with an aluminium plug with flash-hole.

Gaine (Fig. 25)

The gaine has a tubular body of the type used for detonators, with its open end positioned over the flash-hole in the cap of the fuze by two washers, apparently of cork. Details of the gaine filling are not yet available; this probably consists of P.E.T.N./Wax, with a pressing of lead azide and lead styphnate at the mouth.

Case

The case is 9.8 inches in length and has a slight shoulder formed by an increase in the taper. The base is stamped " 3.7 cm. Pak " and with the model number 6331. In some instances the model number is followed by a star. Cases bearing this mark are also of brass. Steel cases coated with brass are also used, and bear the usual stamping, " St.," after the model number. The mouth of the case is attached to the shell by a cannelure.

Propellant Charge and Igniter

The propellant charge consists of a bundle of tubular sticks of double base propellant, with an igniter at the base secured by strips of material to the ties of the bundle. In some instances the propellant consists of diethylene-glycol-dinitrate and nitrocellulose, and in others nitroglycerine and nitrocellulose. The charge weight with the first composition (Digl) is approximately 6 oz. 11 dr., and the diameters of the tube are, external .086 inch and internal .033 inch. With the second composition (Ngl) the charge weight is approximately 6 oz. 4 dr., and the diameters of the tube are, external .098 inch and internal .035 inch. With both compositions the length of the tubes is 7.4 inches.

The igniter bag is stitched to form pockets, and contains approximately 30 grains of granular composition consisting mainly of nitrocellulose, with the addition of potassium nitrate.

The decoppering agent consists of a length of lead (or lead and tin) wire weighing about 77 grains. When the wire is not included the letters " oBD " are stencilled in black on the side of the case.

Primer

A description and drawing of the percussion primer, C/13, is included in this pamphlet.

GERMAN 7.5 CM. PAK 40 CARTRIDGE, Q.F., A.P.C.B.C.
(7.5 cm. Panzergranat-Patrone 39) (Fig. 26)

The round is of the fixed Q.F. type and is used with the 7.5 cm. anti-tank gun, model 40. The overall length is 36.14 inches and the weight 26.5 lb. The shell is painted black and is noticeably different from the earlier type of A.P.C.B.C. shell in that the ballistic cap is smaller and is cannelured to the head of the penetrative cap. The cartridge case is stamped in the base with the designation " 7.5 cm. Pak 44 Rh " or " Pak 44 Rh," but the side of the case is stencilled " 7.5 cm. Pak 40."

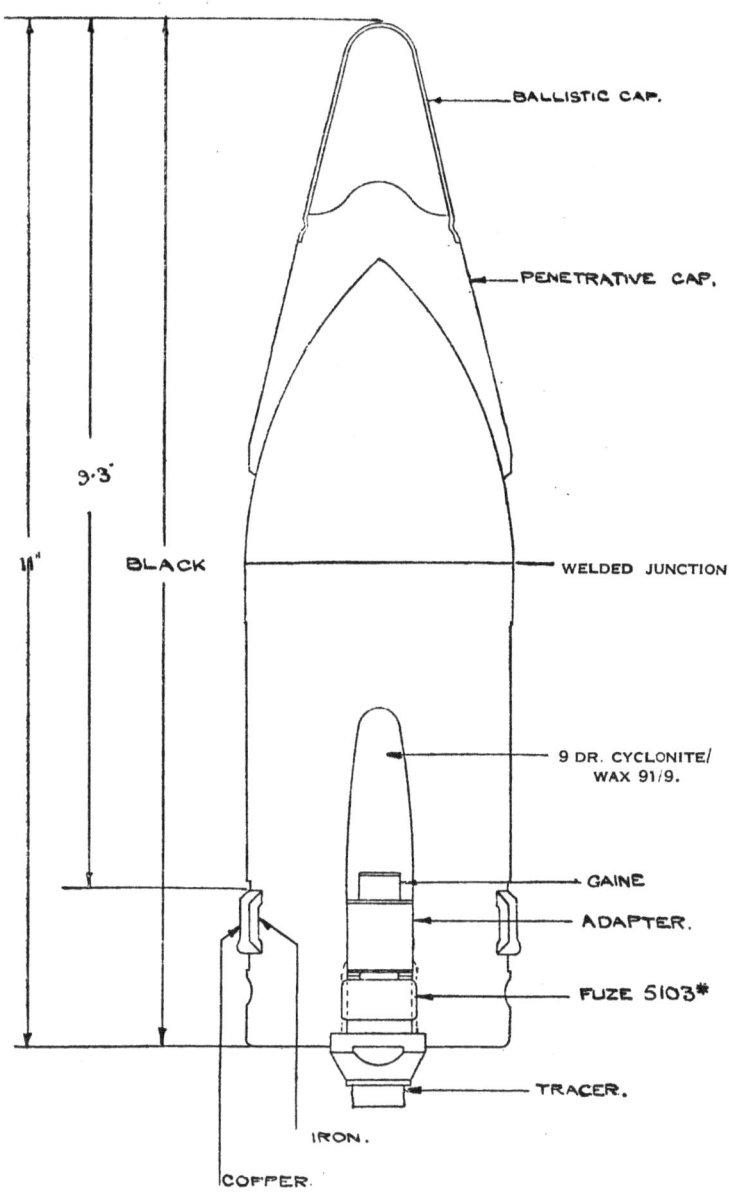

Fig. 26

German 7.5 cm. Pak 40 Kw.K. 40, and Stu K 40, A.P.C.B.C. Shell (Panzergranate 39), with fuze 5103* (Bd.Z. 5103*)

The complete round consists of the following components: —
A.P.C.B.C. shell filled cyclonite/wax (91/9).
Base tracer fuze, model 5103*, with adapter and gaine.
Propellant charge of double base composition with igniter.
Case of steel, coated with brass, model 6340 St.
Primer percussion, model C/12nASt.

Shell (Pzgr. 39)

The shell has a very small cavity for the bursting charge, and is fitted with a driving band of iron covered by copper. A cannelure is formed behind the driving band for the attachment of the case. The shell is closed at the base by the fuze. The penetrative cap with the short ballistic cap cannelured to its head forms a coned head. The bursting charge consists of approximately 9 drams of cyclonite/wax (91/9), and is provided with a short cavity to receive the gaine. The weight of the shell, filled and fuzed, is approximately 15 lb. Metallurgical details are not yet available. The weight of the empty shell without the fuze is 14 lb. 11 oz. 13 dr.

Tracer Fuze and Gaine

A description of the fuze 5103* with the gaine and adapter used is given in Pamphlet No. 4 on pages 14 and 15. The tracer is similar to that shown in Fig. 25 in this pamphlet.

Propellant Charge

The propellant charge has a weight of approximately 6 lb. and consists of tubular sticks 24.6 inches in length, with diameter .14 inch external and .05 inch internal. The charge is covered from the base for about half the length by a bag which carries an igniter at the bottom. The propellant is of the double base type, and consists basically of nitrocellulose and diethylene-glycol-dinitrate. The bag is marked " 7.5 cm. Pak 40, 2.750 kg. Digl. R.P.-G1, 625-3, 8/1, 3."
Details of the igniter composition are not yet available.

Case

The case is of steel, coated with brass, and is 26.2 inches in length. The base is stamped with the model number " 6340 St " and " Pak 44 Rh " or " 7.5 cm Pak 44 Rh " instead of Pak 40. This model number and the designation " Pak 44 Rh," with calibre omitted, has been met with on the case used with 7.62 cm. Pak 36 rounds, a case 28 inches in length. Apparently the case was designed for a Pak 44 equipment, but is being utilized for the converted 7.62 cm. Pak 36 and the 7.5 cm. Pak 40 equipments. The details of the charge as marked on the cartridge bag are stencilled on the side of the case.

Primer

The percussion primer, C/12nA, is described in Pamphlet No. 4, page 10. The letters " St " added to the designation indicate that the primer is of steel.

GERMAN 7.5 CM. PAK 40 CARTRIDGE, Q.F., H.E.
(7.5 cm. Sprenggranat-Patrone 34)

The fixed Q.F. round is used with the 7.5 cm. anti-tank gun, model 40. The overall length is 37.36 inches and the weight 20.25 lb. The shell is fitted with a fuze of the A.Z.23 type, and is painted a deep olive green. The numeral "13" is stencilled in black above the shoulder of the shell and indicates a bursting charge of amatol 40/60. The base of the case is stamped with the model number "6340 St" and with the designation "Pak 44 Rh," but the side of the case is stencilled "7.5 cm. Pak 40."

The complete round consists of the following components:—
H.E. shell filled amatol 40/60 with smoke box.
Fuze Kl.A.Z.23.
Gaine, Gr. Zdlg C/98.
Propellant charge of double base composition with igniter.
Case of steel, coated with brass, model 6340 St.
Primer percussion, model C/12nASt.

Shell (Sprgr. 34)

The shell, filled and fused, weighs 12.8 lb. The driving band may be of the copper covered iron type, or may be wholly of iron. A cannelure is formed behind the driving band for the attachment of the case. The wall of the shell is tapered, being thicker towards the base than towards the head, thus increasing the capacity for the bursting charge. A fuze-hole bush is screwed in at the nose to take the fuze. The bush is fitted with an exploder container.

The method of filling design is the same as that for the 7.5 cm. H.E. shell described on page 32 and shown in Pamphlet No. 4 (Fig. 21), where details of the smoke box (Rauchentwickler No. 8) and the gaine are included. The weight of the filling is 1 lb. 6 oz. 12 dr.

Fuze

The Kl.A.Z.23, small type of the model 23 fuze, is similar in construction and action to the A.Z.23 described in Pamphlet No. 1, Section 4.

Propellant Charge

The 1 lb. 11½ oz. charge is contained in an artificial silk bag with an igniter sewn at the base, and is similar to that for the 5 cm. cartridge described in Pamphlet No. 4, pages 9 and 10, *i.e.*, the bag containing flake propellant with a long central tubular stick of propellant which protrudes through the choked neck of the bag and reaches almost to the base of the shell. An annular bag containing potassium sulphate is secured at the choke of the cartridge bag and a ring pushed down over the neck of the bag keeps the flake propellant towards the base of the case. The flake propellant is of the "Gu.Bl.P.-AO-(4.4.0,6)" variety, which consists of diethylene-glycol-dinitrate, nitro-guanidine and nitro-cellulose. The dimensions of the flake are .16 x .16 x .02 inches. Details of the tubular propellant and the igniter composition are not yet available.

The cartridge bag is marked "7.5 cm. Pak 40, 780g, Gu.Bl.P.-AO-(4.4.0,6)."

Case and Primer

These are the same as those described for the A.P.C.B.C. cartridge in this pamphlet.

GERMAN 7.5 CM. Kw.K. 40 and Stu K. 40 CARTRIDGE, Q.F., A.P.C.B.C. (7.5 cm. Kw.K. 40 u Stu K. 40 Panzergranat-Patrone 39)

This fixed Q.F. round is used with the 7.5 cm. tank gun, model 40, and the 7.5 cm. assault gun, model 40. The overall length of the round is 29.3 inches and the weight 24.8 lb. The shell with its penetrative and small ballistic cap is painted black, and is the same as that described for the 7.5 cm. Pak 40 in this pamphlet. The necked cartridge case is 19.4 inches long and is fitted with the electric primer, C/22. The model number of the case, " 6339 St," and the designation of the tank equipment, " 7.5 cm. Kw.K. 40," are stamped in the base of the case.

Propellant Charge

The propellant charge is contained in an artificial silk bag which bears the markings " Für Trop. P.T. + 25° C., 7.5 cm. Stu K. 40, 7.5 cm. Kw.K. 40, 2160 Kg. Digl. R.P. - G.1 - $\frac{370}{420}$ 3.8/1.5." An igniter is formed at the base of the bag.

The propellant is in the form of tubular cords 14.6 inches and 16.5 inches in length. These lengths are indicated in millimetres by the figures " 370 " and " 420 " marked on the bag and case. The external and internal diameters of the tube are .152 − .053 inches. The double base propellant consists basically of 59.81 per cent. of nitrocellulose (nitrogen content 11.7 per cent.) and 27.07 diethylene-glycol-dinitrate, and includes centralite, potassium sulphate, graphite and a waxy material. The weight of the charge as indicated by the markings for a charge temperature normal of 25° C. is 4 lb. 12 oz. 2 dr.

The igniter filling is in the form of chopped cord which is porous and green in colour. The composition as found by analysis consists of: 93.10 per cent. of nitrocellulose (nitrogen content 13.2 per cent.), .65 per cent. of diphenylamine and .7 per cent. of potassium sulphate. Graphite and oily material are also included. The weight of the igniter filling is 315 grains.

GERMAN 7.5 CM. Kw.K. 40 and Stu K. 40 CARTRIDGE, Q.F., H.E. (7.5 cm. Kw.K. 40 u Stu K. 40 Sprenggranat-Patrone)

This fixed Q.F. round is used with the 7.5 cm. tank gun, model 40, and the 7.5 cm. assault gun, model 40. The overall length of the round is 30.6 inches and the weight is approximately 20 lb. The shell is fitted with the Kl.A.Z.23 fuze and painted a dark olive green with the numeral " 13 " stencilled in black above the shoulder. This shell is the same as that described for the 7.5 cm. Pak 40 in this pamphlet. The necked cartridge case is 19.4 inches long and is fitted with the electric primer, C/22. The model number of the case, " 6339 St." and the designation of the tank equipment, " 7.5 cm. Kw.K. 40," are stamped in the base of the case.

Propellant Charge

The propellant charge is contained in an artificial silk bag which bears the markings " Für Tropen P.T.+25° C., 7.5 cm. Stu K. 40, 7.5 cm. Kw.K. 40, 745 g., Gu.Bl.P.–AO–(4.4.0,6)." The weight of the charge as indicated by the markings is 1 lb. 10 oz. 4 dr.

The charge consists of square graphited flake and a large stick of multi-tubular propellant which protrudes from the choked neck of the bag and reaches almost to the base of the shell. A small circular annular bag, marked " 20 g. K_2SO_4," is fitted over the tubular propellant at the neck of the main bag. The annular bag contains 278 grains of potassium sulphate.

The dimensions of the flake propellant are $.156 \times .156 \times .025$ inches. The composition includes nitrocellulose, diethylene-glycol-dinitrate, nitroguanidine and graphite. The stick of multi-tubular propellant is black and has a diameter of .554 inch with a cross-web section and four quadrant holes. The web measures .042 inch. The propellant is of the double base type ; details of the composition are not yet available.

The igniter formed at the base of the main charge bag contains 335 grains of nitrocellulose powder in the form of chopped cord. The powder is porous and is greyish-green in colour.

GERMAN 10.5 CM. l.F.H.18 Q.F. CARTRIDGES

The cartridge used in the 10.5 cm. l.F.H.18 (gun-howitzer) is of the separate loading type, and consists of a short case containing 5 charges. The case is closed at the mouth by a cardboard cup and is fitted with the percussion primer C/12nA. A special charge, section 6, is also provided for use in the case in place of sections 1 to 5 as a super charge.

Cases

Two types of these cases have been met with, both of which are 6.1 inches in length and increase in taper near the mouth. The model numbers of the cases are 6342/65 and 6342. The 6342/65 model is of steel coated with brass.

Model 6342/65 is of the built-up type, the wall being turned in at its base end to seat in a concave seating formed in the front face of the base, where it is secured by a steel disc which is held by a steel circular nut screwed to the primer boss. Both the disc and the nut are brass plated. Four circular recesses are formed in the base near the flange. These are apparently used in assembling the case.

Model 6342 is the normal type of solid drawn case, and is of brass.

Propellant Charge (Sections 1 to 5)

The propellant charge is made up in five sections. The sections are contained in flat circular bags of silk or artificial silk which are marked in black to indicate the charge numeral, the details of the propellant, and the place and date of filling. An igniter of nitrocellulose powder is sewn to the base of the section No. 1.

The weight and nature of propellant charges met with are as follows:
 1 lb. 5 oz. Ngl.Bl.P.-12.5 (double base nitroglycerine/nitrocellulose).
 1 lb. 8 oz. Digl.Bl.P.-10.5 (double base diethylene-glycol-dinitrate/
  nitrocellulose).

The latter charge weight is for a charge temperature normal of 25° C., whereas the former is for a 10° C. normal. The greater weight of the Digl. Bl.P.-10.5 charge, even when reduced for use in hot climates, indicates the superiority of the Ngl. Bl.P.-12.5 propellant.
The weight and size of the propellant in the sections for the two natures are:

Ngl. Bl.P.-12.5 charge.

Section	Size of Flake		Weight of Section		
	Mm.	Inches	Grams	Oz.	Dr.
1	4 × 4 × ·02	·16 × ·16 × ·008	180	6	6
2	4 × 4 × 1	·16 × ·16 × ·04	55	1	15
3	4 × 4 × 1	·16 × ·16 × ·04	60	2	2
4	4 × 4 × 1	·16 × ·16 × ·04	115	4	1
5	4 × 4 × 1	·16 × ·16 × ·04	185	6	8

Digl. Bl.P.-10.5 charge. (Normal charge temperature 25° C.)

Section	Size of Flake		Weight of Section		
	Mm.	Inches	Grams	Oz.	Dr.
1	3 × 3 × 0·8	·12 × ·12 × ·03	241	8	8
2	3 × 3 × 0·8	·12 × ·12 × ·03	59	2	1
3	3 × 3 × 0·8	·12 × ·12 × ·03	70	2	8
4	3 × 3 × 0·8	·12 × ·12 × ·03	118	4	3
5	3 × 3 × 0·8	·12 × ·12 × ·03	192	6	12

Special Charge (Sonderkart 6)

The special charge, Section 6, is issued packed in a rolled paper cylindrical container and is inserted in the case when required for use in place of the normal contents.

The charge is contained in a white artificial silk bag which is choked at the neck and has an igniter stitched to the base. The propellant is in flake form, .16 × .16 × .047 inches. The charge weight, adjusted for a standard charge temperature of 25° C., with the double base propellant consisting basically of diethylene-glycol-dinitrate and nitrocellulose, is 1 lb. 11 oz. 10 dr. The bag is marked with the charge numeral "6D," the designation of the equipment "l.F.H.18," and the weight, nature, shape and size of the propellant "784 g. Digl. Bl.P.10.5 (4.4.1,2)."

The igniter contains 7 oz. 13 dr. of propellant flake of the same composition as the propellant charge, but of a smaller size.

The package containing the charge is stencilled "Sonderkart 6 der l.F.H.18" and carries an instructional label to the effect that before inserting the special charge in the cartridge case, the other sectional charges including the section carrying the igniter should be removed.

www.ingramcontent.com/pod-product-compliance
Lightning Source LLC
Chambersburg PA
CBHW032012080426
42735CB00007B/583